All the Math You'll Ever Need

Wiley Self-Teaching Guides teach practical skills in mathematics and science. Look for them at your local bookstore.

Other Science and Math Wiley Self-Teaching Guides:

Science

Basic Physics: A Self-Teaching Guide, Third Edition, by Karl F. Kuhn and Frank Noschese

Biology: A Self-Teaching Guide, Third Edition, by Steven D. Garber

Chemistry: A Self-Teaching Guide, Third Edition, by Clifford C. Houk, Richard Post, and Chad A. Snyder

Math

All the Math You'll Ever Need: A Self-Teaching Guide, by Steve Slavin

Practical Algebra: A Self-Teaching Guide, Second Edition, by Peter H. Selby and Steve Slavin

Quick Algebra Review: A Self-Teaching Guide, by Peter H. Selby and Steve Slavin

Quick Business Math: A Self-Teaching Guide, by Steve Slavin

Quick Calculus: A Self-Teaching Guide, Second Edition, by Daniel Kleppner and Norman Ramsey

All the Math You'll Ever Need

A Self-Teaching Guide

Third Edition

Carolyn C. Wheater
Steve Slavin

JB JOSSEY-BASS™

A Wiley Brand

Jossey-Bass
A Wiley Imprint
111 River St, Hoboken, NJ 07030
www.josseybass.com

Jossey-Bass books and products are available through most bookstores. To contact Jossey-Bass directly, call our Customer Care Department within the U.S. at 800-956-7739, outside the U.S. at +1 317 572 3986, or fax +1 317 572 4002.

Wiley also publishes its books in a variety of electronic formats and by print-on-demand. Some material included with standard print versions of this book may not be included in e-books or in print-on-demand. If this book refers to media such as a CD or DVD that is not included in the version you purchased, you may download this material at http://booksupport.wiley.com. For more information about Wiley products, visit www.wiley.com.

Library of Congress Cataloging-in-Publication Data is Available:

ISBN 9781119719182 (paperback)
ISBN 9781119719229 (epdf)
ISBN 9781119719199 (ebook)

Cover Design: Paul McCarthy
Cover Image: © Frankramspott/Getty Images

SKY10033204_031122

Contents

How to Use This Book

This book is organized by chapter with periodic self-tests throughout each chapter. Their purpose is to make sure you comprehend material before moving on. If you find that you have made an error, look back at the preceding material to make sure you understand the correct answer. The information is arranged so that it builds on what comes before. To fully understand the information at the end of a chapter, you must first have completed all of the preceding self-tests.

The format of this book lends itself to proper pacing. When you're going too slowly, you'll say to yourself, "This stuff is so easy—I'm getting bored." You'll be able to skip a few sections and move on to new material. But when you find yourself pounding your fists against the wall and despairing of ever learning math, that may mean you've been moving ahead a bit too quickly.

If you feel that you don't need to read a particular chapter, you may want to take the self-tests anyway. These provide not only a quick review of the subject matter covered in the chapter, but also a good way of gauging what you already know.

Should you find, on the other hand, that you're having trouble doing a certain type of problem, it will be made clear to you that you need to review an earlier section. For example, no one can do simple division without knowing the multiplication table, so everyone who gets stuck at this point will be sent back to learn that table once and for all. Once that's accomplished, it will be clear sailing through the next few chapters.

This book provides a fast-paced review of arithmetic and elementary algebra, with a smattering of statistics thrown in. It is intended to refresh the memory of the high school or college graduate.

The main emphasis here is on getting you to rely on your own mathematical skills. No longer will you be intimidated trying to calculate tips. No longer will you need to whip out your pocket calculator to do simple arithmetic. And you won't have to wait months to see tangible results. You won't even have to wait weeks. In just a few days your friends and colleagues will notice your new mathematical muscles. So don't delay another minute. Turn to Chapter 2 and just watch those brain cells start to grow.

Acknowledgments

From Steve: Many thanks are due, so I'd like to name names. My longtime editor at Wiley, Judith McCarthy, made hundreds of suggestions to improve and update the book. Authors often hate to change even a word, but Judith's editing has made this second edition a much smoother read. Claire McKean did a thorough copyedit, catching dozens of errors that made it through the first edition, and Benjamin Hamilton supervised the production of the book from copyediting through page proofs.

I owe a large debt of gratitude to my family, especially to my nephews, Jonah and Eric Zimiles. Jonah provided me with a blow-by-blow critique of the strengths and weaknesses of my previous book, *Economics: A Self-Teaching Guide* (Wiley, 1988), on which I was able to build while writing *this* book. And Eric, after having read that book, recognized its format lent itself best to my writing style and encouraged me to write another book. Eric's daughters, Eleni, 11, Justine, 7, and Sophie, 5, have contributed to the new edition by helping me with my math whenever I happened to get stuck.

My father, Jack, a retired math teacher, provided inspiration of another kind. As the oldest living academic perfectionist, he upholds such unattainable standards that one cannot help but feel tolerance for one's own shortcomings and those of just about everyone else. And finally, I wish to thank my sister, Leontine Temsky, for her rationality and common sense in the most uncommon and irrational of times.

From Carolyn: Every opportunity to work with the incredible folks at Wiley has been a pleasure. This project was no exception. My thanks to go Christine O'Connor, Tom Dinse, and Riley Harding for providing the vision for the project, giving me the freedom to make it my own, and guiding me through every step. I'm grateful also to copyeditor Julie Kerr, whose keen eye and infinite patience make the final product so much better.

All the Math You'll Ever Need

1 Getting Started

Far too many Americans are mathematically illiterate. Although many of these people are college graduates, they have trouble doing simple arithmetic. One cannot help but wonder how so many people managed to get so far without having mastered basic arithmetic. Math phobia seems to have become fashionable. People who would never think it amusing to claim not to be able to read or write chuckle as they announce, "I can't do math."

We all have to deal with numbers *sometime*—in banking, on taxes, in choosing a mortgage. Like it or not, numbers are an important part of our lives, and the importance of numerical literacy is increasing in finance, economics, science, government, and more. It is time math stopped intimidating us.

What we'll be doing in this book is going back to basics. We'll focus on the multiplication table. You'll need to memorize it. If you need an even more basic text, you can refer to one such as *Quick Arithmetic: A Self-Teaching Guide, 3rd edition* by Robert A. Carmen and Marilyn J. Carmen (Wiley, 2001).

In *All the Math You'll Ever Need*, the use of complex formulas is generally avoided. Although such formulas have an honored place in mathematics, they rarely need to be memorized. The ones that are used frequently work their way into memory. The others can be looked up when they're needed.

Finally, the use of technical terms is minimized whenever possible. Having the vocabulary to describe mathematical ideas and operations accurately is important to learning but you don't need a lot of fancy language for that. There are no quadratic formulas, logarithmic tables, integrals, or derivatives, and there are only a handful of very simple graphs.

This book was designed to be explored without ever using a calculator or computer. Don't get nervous. You will not be asked to throw away your calculator. Just put it in a safe place for now, to be taken out and used only on proper occasions. A calculator is most effectively used for three tasks: (1) to do calculations that need to be done rapidly, (2) to do repetitive calculations, and (3) to do sophisticated calculations that would take a great deal of time to do without a calculator. Calculators and computers are fast and as accurate as their users allow them to be. Typos are a thing, even on calculators. You need to first know what you want to ask the calculator to do, and then have enough math knowledge to decide if the answer it gives you makes sense.

The trick is to use our calculators for these specific tasks and not for arithmetic functions that we can do in our heads. So put away your calculator and start using your innate mathematical ability.

2 Essential Arithmetic

Every number system (and, yes, there are or have been others) is made up of a set of symbols that we call numbers and one or more operations you can perform with them. Those operations make up what we call *arithmetic*. The basic operation in our number system is addition, the act of putting together. The other operations—multiplication, subtraction, division—are related to, or built from, addition.

1 ADDITION

Addition is, at its heart, about counting. If you have 6 pair of shoes and you buy 3 new pairs, counting will tell you that you now have 9 pairs. You added 6 + 3 and got an answer of 9 by counting. After a while you don't have to count every time, because you get to know that 6 + 3 = 9.

You store a lot of addition facts like that in your memory, but there's a limit to how much memorization can help. You probably know that 4 + 8 = 12, but you're unlikely to memorize the answer to 5,387 + 9,748. Adding larger numbers requires a little more information about our number system.

Place Value

Our number system is a place value system, meaning that the value of a numeral depends on the place it sits in. In the number 444 each 4 has a different meaning. The 4 on the right is in the ones place so it represents 4 ones or simply 4. The 4 on the left is in the hundreds place and represents 4 hundreds or 400. The middle 4 is in the tens place so it represents 4 tens or 40. The number 444 is a shorthand for 400 + 40 + 4.

That expanded form, 400 + 40 + 4, helps to explain how we add large numbers. We add the ones to the ones, the tens to the tens, the hundreds to the hundreds and on up in the place value system. If you need to add 444 + 312, think:

$$400 + 40 + 4$$
$$+ \ 300 + 10 + 2$$

Add the 4 ones and the 2 ones to get 6 ones, the 4 tens with 1 ten to get 5 tens and the 4 hundreds with 3 hundreds to get 7 hundreds. Now that would look like this:

$$400 + 40 + 4$$
$$300 + 10 + 2$$
$$\overline{700 + 50 + 6}$$

You're probably thinking that you could just write the numbers underneath one another in standard form and add down the columns, and you'd be absolutely correct.

$$
\begin{array}{r}
444 \\
+\ 312 \\
\hline
756
\end{array}
$$

The reason to think about it in expanded form, at least for a few minutes, comes up when you have to add something like 756 + 968. The basic rule is the same.

$$
\begin{array}{l}
7 \text{ hundreds} +\ 5 \text{ tens} + 6 \text{ ones} \\
9 \text{ hundreds} +\ 6 \text{ tens} + 8 \text{ ones} \\
\hline
16 \text{ hundreds} + 11 \text{ tens} + 14 \text{ ones}
\end{array}
$$

But you can't squeeze 16 (or 11 or 14) into one place. 756 + 968 does not equal 161114. You've got to do some regrouping, or what's commonly called *carrying*. Those 14 ones equal 1 ten and 4 ones. You're going to keep the 4 ones in the ones place and move the ten over to the middle place with the rest of the tens. That will turn

$$
\begin{array}{l}
7 \text{ hundreds} +\ 5 \text{ tens} +\ 6 \text{ ones} \\
9 \text{ hundreds} +\ 6 \text{ tens} +\ 8 \text{ ones} \\
\hline
16 \text{ hundreds} + 11 \text{ tens} + 14 \text{ ones}
\end{array}
$$
into
$$
\begin{array}{l}
7 \text{ hundreds} + \overset{1 \text{ ten}}{\underset{5 \text{ tens}}{}} + 6 \text{ ones} \\
9 \text{ hundreds} + 6 \text{ tens} + 8 \text{ ones} \\
\hline
16 \text{ hundreds} + 12 \text{ tens} + 4 \text{ ones}
\end{array}
$$

You'll do the same sort of regrouping with the 12 tens. Ten of those tens make 1 hundred, leaving 2 tens in the tens place. You can do this without using the expanded form. Add 6 + 8 to get 14. Put down the 4 and carry the one ten.

$$
\begin{array}{r}
{}^{1} \\
756 \\
+\ 968 \\
\hline
4
\end{array}
$$

Add $1 + 5 + 6$ to get 12. Put down the 2 (tens) and carry the 1 (hundred).

$$
\begin{array}{r}
{\scriptstyle 1\,1} \\
756 \\
+\ 968 \\
\hline
24
\end{array}
$$

Add $1 + 7 + 9$ to get 17. The 7 goes in the hundreds place and the 1 (thousand) slides into the thousands place.

$$
\begin{array}{r}
{\scriptstyle 1\,1} \\
756 \\
+\ 968 \\
\hline
1{,}724
\end{array}
$$

Problem 1:

Add 312 and 423.

Solution:

$$
\begin{array}{r}
312 \\
+\ 423 \\
\hline
735
\end{array}
$$

All that's necessary is adding the digits in each column: $2 + 3 = 5$, $1 + 2 = 3$, and $3 + 4 = 7$.

Problem 2:

What is the result when 459 is added to 1,276?

Solution:

$$
\begin{array}{r}
{\scriptstyle 1\,1} \\
1{,}276 \\
+\ 459 \\
\hline
1{,}735
\end{array}
$$

This one requires a little bit of regrouping. Add $6 + 9$ to get 15, put down the 5 and carry 1 to the next column. Then $7 + 5$ is 12, plus the 1 you carried is 13. Put down the 3 and carry the 1. You can think of the rest as $2 + 4 + 1 = 7$ and the 1 thousand comes down unchanged, or you can think of it as $12 + 4$ is 16, plus 1 you carried is 17.

Problem 3:

What is the combined total of 9,671 and 2,859?

Solution:

$$
\begin{array}{r}
{\scriptstyle 1\ \ 11} \\
9,671 \\
+\,2,859 \\
\hline
12,530
\end{array}
$$

Here again you're regrouping. In the ones column, 1 + 9 is 10, so put down the 0 and carry the 1. Then 7 + 5 is 12 plus 1 you carried makes 13. Put down the 3 and carry the 1. Add 6 + 8 + 1 to get 15. Put down the 5 and carry the 1. Finally, 9 + 2 + 1 is 12.

2 MULTIPLICATION

Multiplication is repeated addition. For instance, you probably know 4 × 3 is 12 because you searched your memory for that multiplication fact. There's nothing wrong with that.

Another way to calculate 4 × 3 is to think of it as adding four threes, or adding three fours.

$$3 + 3 + 3 + 3 = 12 \qquad \text{or} \qquad 4 + 4 + 4 = 12$$

What about 5 × 7? Maybe you know it's 35, but you could always do this:

$$7 + 7 + 7 + 7 + 7 = 35 \qquad \text{or} \qquad 5 + 5 + 5 + 5 + 5 + 5 + 5 = 35$$

You do multiplication instead of addition because it's shorter—sometimes much shorter. Suppose you needed to multiply 78 × 95. If you set this up as an addition problem, you'd have to write 78 copies of 95 before you could even start adding.

Let's set this up as a regular multiplication problem and take a look at the expanded form.

$$
\begin{array}{r}
95 \\
\times\ 78
\end{array}
\quad \text{becomes} \quad
\begin{array}{r}
90 + 5 \\
\times\ 70 + 8
\end{array}
$$

The key to this multiplication is you have to multiply 8 × 5 and 8 × 90 and then multiply 70 × 5 and 70 × 90, and add up all the results. Don't get discouraged, because there is a condensed form.

The first set of numbers we'd multiply would be 8 × 5. You probably know, or can figure out, that's 40. (We'll focus on all the multiplication facts you should memorize in Chapter 3, "Focus on Multiplication.") Then we'd multiply 8 × 90, which just means multiplying 8 × 9 and putting a zero at the end. Whenever you multiply a number that ends in zero, you can deal with the non-zero parts and add the zero at the end. (See Chapter 5, "Mental Math" for more on that shortcut.) 8 × 9 = 72 so 8 × 90 = 720. Next would come 70 × 5. 7 × 5 = 35 so 70 × 5 = 350. The last multiplication would be 70 × 90. Multiply 7 × 9 = 63, and then add a zero for the 70 and another zero for the 90. 70 × 90 = 6,300. Add up 6,300 + 350 + 720 + 40 to get 7,410.

$$
\begin{array}{r}
95 \\
\times\,78 \\
\hline
40 \\
720 \\
350 \\
6300 \\
\hline
7410
\end{array}
$$

Here's how to write it more compactly. Multiply 8 × 5 = 40, put down the 0 and carry the 4. 8 × 9 = 72 and the 4 we carried makes 76. Write the 76 in front of that 0 you put down and you see 760. This 760 is the 40 and the 720 combined. Now, you need to multiply 95 by 70, which means multiply by 7 and add a zero. So put the zero down first, under the 0 of the 760. Then 7 × 5 = 35. Put down the 5 to the left of the 0 and carry the 3. 7 × 9 = 63 plus the 3 you carried is 66. Write the 66 in front of the 50 and you've got 6,650, which is the 350 and 6300 combined. Add the two lines, and you're done.

$$
\begin{array}{cccc}
\overset{4}{95.} & 95 & \overset{3}{95} & \overset{3}{95} \\
\times 78 & \times 78 & \times 78 & \times 78 \\
\hline
760 & 760 & 760 & 760 \\
 & 0 & 50 & 6650 \\
 & & & \hline
 & & & 7410
\end{array}
$$

As you can see, a long multiplication problem can be broken down into a series of simple multiplication problems. It's important to have basic multiplication facts in memory, so you don't have to spend time doing the repeated addition every time. You'll learn more about that in the next chapter.

Problem 1:

Multiply 73 by 5.

Solution:

$$
\begin{array}{r}
^{1} \\
73 \\
\times 5 \\
\hline
365
\end{array}
$$

Begin with 5 × 3 = 15. Put down the 5 and carry 1. Then 5 × 7 is 35 plus the 1 you carried is 36.

Problem 2:

Find the product of 86 and 12.

Solution:

First multiply 86 by 2.

$$
\begin{array}{r}
^{1} \\
86 \\
\times 12 \\
\hline
172
\end{array}
$$

Multiplying 2 × 6 gives you 12, so put down the 2 and carry 1. Then 2 × 8 is 16 plus 1 you carried is 17.

Place a zero at the end of the second line, or if you prefer, just move one space right, and multiply 1 × 86, which obviously is 86. Add the columns to complete the job.

$$
\begin{array}{r}
86 \\
\times 12 \\
\hline
^{1} \\
172 \\
860 \\
\hline
1{,}032
\end{array}
$$

Problem 3:

What is the result when 125 is multiplied by 32?

Solution:

Multiplying 125 by 2 requires a little bit of carrying.

$$
\begin{array}{r}
{\scriptstyle 1} \\
125 \\
\times\, 32 \\
\hline
250
\end{array}
$$

Place a zero on the second line, or move one space right, then multiply 3×125. Add down each column, regrouping where necessary.

$$
\begin{array}{r}
{\scriptstyle 1} \\
125 \\
\times 32 \\
\hline
{\scriptstyle 1} \\
250 \\
{\scriptstyle 1} \\
3750 \\
\hline
4{,}000
\end{array}
$$

Ready to test yourself? Try Self-Test 2.1.

SELF-TEST 2.1

1. Add 453 and 975.

2. Find the sum of 1,864 and 798.

3. Multiply 561 by 92.

4. What is the product of 891 and 30?

5. Multiply 135×112.

If the first two gave you trouble, review Frame 1. If you got any of the last three wrong, review Frame 2. If you've got this, move on!

3 SUBTRACTION

Subtraction is the inverse, or opposite, of addition. Addition puts together. Subtraction takes apart. If you buy a carton of 12 eggs and you use 4 of them to make breakfast, how many eggs are left? $12 - 4 = 8$ if you count the remaining eggs.

That subtraction problem is another way of thinking about the addition problem $4 + 8 = 12$. Each subtraction problem has a related addition problem. $17 - 9 = ?$ is related to $9 +$ what $= 17$? Sometimes it's easier to think about the addition version. If you have 9 and you want 17, you need 1 to make 10 and then 7 more to get up to 17. That's a total of 8.

For larger problems, you'll work column by column, starting from the ones, just like addition, but sometimes you'll need to "borrow," which means you'll regroup but in the other direction.

In the following example, the ones column is easy: $8 - 3 = 5$. But trying to take 9 away from 3 in the tens column is a problem. If you only have 3, how can you take away 9? (There's more than one answer to that question. We'll look at one now and another one in Chapter 6, "Positive and Negative Numbers.")

$$\begin{array}{r} 638 \\ -193 \\ \hline ??5 \end{array}$$

We only have 3 tens in our tens column, and we need more, so we're going to borrow 1 hundred from the hundreds column and exchange it for 10 tens. There are 6 hundreds, so if we borrow 1, there will be 5 left. We'll exchange the 1 hundred for 10 tens and add them to our 3 tens so we have 13 tens. We can subtract 9 tens from 13 tens and get 4 tens, then subtract 1 hundred from the remaining 5 hundreds to get 4 hundreds.

$$\begin{array}{r} \overset{5}{\cancel{6}}{}^{1}38 \\ -193 \\ \hline 445 \end{array}$$

Problem 1:

Subtract 124 from 896.

Solution:

$$\begin{array}{r} 896 \\ -124 \\ \hline 772 \end{array}$$

This subtraction doesn't require any regrouping. Just subtract the digits in each column: $6 - 4 = 2$, $9 - 2 = 7$, and $8 - 1 = 7$.

Problem 2:

Find the difference between 293 and 581.

Solution:

To begin, you need to "borrow" and regroup, because you can't subtract 3 from 1. Remember that the 8 tens become 7 tens and the ten we borrowed is added to the 1. Subtract 3 from 11 to get 8.

$$\begin{array}{r} {\scriptstyle 7} \\ 5\cancel{8}{}^{1}1 \\ -2\,9\,3 \\ \hline 8 \end{array}$$

But this requires another regrouping. Borrow 1 from the 5 hundreds, leaving 4 hundreds, and change the borrowed hundred to 10 tens. Add that to the 7 (not 8) tens and subtract 9 from 17.

$$\begin{array}{r} {\scriptstyle 4}{}^{1}{\scriptstyle 7} \\ \cancel{5}\cancel{8}{}^{1}1 \\ -2\,9\,3 \\ \hline 2\,8\,8 \end{array}$$

Finish by subtracting 2 hundred from the remaining 4 hundred.

4 DIVISION

Division is the opposite, or inverse, of multiplication. You can think of a division question like $21 \div 3 = ?$ as the related multiplication question $3 \times$ what $= 21$? Sometimes memory is enough to answer that. You can also phrase the question as, "If you put 21 objects into groups of 3, how many groups can you make?" Another approach is to ask, "If 3 people share $21, how much does each person get?"

Division with larger numbers can get complicated. There are several methods, one of which is covered below. Others will come up in Chapter 4, "Focus on Division." It's fair to say that division with larger numbers is one task for which calculators are helpful. If you're going to use a calculator, you'll want to be able to estimate the answer, so that you can see if the answer on the calculator is reasonable. Errors, even as simple as a typing error, can give you a wildly wrong answer.

Estimation

When you use a calculator for division, it helps to have an idea of what a reasonable answer will look like. If you divide 3,482,603 by 274 and get an answer of 12, bells should go off to say that can't possibly be right. Why doesn't that sound right? Well, think about dividing with easier numbers. 3,482,603 is larger than 3,000,000 and 274 is around 300. $3,000,000 \div 300 = 10,000$ so the answer to the actual problem should be in that neighborhood, which means 12 must be wrong. In fact, $3,482,603 \div 274 = 12,710$ with a little left over.

Short Division

In a division problem, you have a dividend, a divisor, a quotient, and a remainder. You don't need those labels all that often but we'll need them to talk about how to divide. If you want to divide 785 by 5, the 785 is the dividend and 5 is the divisor. If the dividend is large but the divisor is small, you can use what's called short division. It's a method that uses regrouping and compresses the work.

Set up with the dividend inside and the divisor outside, like this: $5\overline{)785}$. Look at the first digit in the dividend, 7. Can you divide 7 by 5? There is one 5 and 2 left over. Show that like this:

$$\frac{1}{5\overline{)7^285}}$$

Those 2 hundreds that were left get regrouped as 20 tens so now that middle number is not just 8 but 28. Divide 28 by 5, and you get 5 with 3 left over.

$$\frac{1\ 5}{5\overline{)7^28^35}}$$

Divide 35 by 5 to get 7.

$$\frac{1\ 5\ 7}{5\overline{)7^28^35}}$$

Problem 1:
What is the result of dividing 161 by 7?

Solution:

$$\frac{23}{7\overline{)16^21}}$$

Divide 7 into 16. It goes twice, so place the 2 up top, but $7 \times 2 = 14$, so there will be 2 left over. Write the 2 beside the 1 and divide 7 into 21. That gives you 3.

Problem 2:

Divide 1,829 by 31. Estimate first by thinking about 1,800 ÷ 30.

Solution:

Estimate your answer by thinking about how many times 30 could be subtracted from 1,800 or by thinking that $18 \div 3 = 6$ and so $1,800 \div 30$ would need a 6 and another zero. $1,800 \div 30 = 60$, so expect a quotient close to 60. This one may be hard to keep all in your head, so write down as you go.

You're probably tempted to start with a 6, but if you check $6 \times 31 = 186$, and you're trying to divide into 182, so 6 is too big. Drop down to 5, and place $5 \times 31 = 155$ under the 182.

$$
\begin{array}{r}
5 \\
31 \overline{)\,1,829} \\
155
\end{array}
$$

Subtract $182 - 155 = 27$ and bring down the 9. Three goes into 27 nine times so there's a good chance 31 goes into 279 nine times. Multiply to check it.

$$
\begin{array}{r}
5\,9 \\
31 \overline{)\,1,829} \\
155 \\
\hline
27\,9 \\
279
\end{array}
$$

If this longer division is giving you trouble, there's more on this in Chapter 4, "Focus on Division." In the meantime, try Self-Test 2.2.

SELF-TEST 2.2

1. Subtract 384 from 497.

2. What is the difference between 2,857 and 6,793?

3. Divide 4,816 by 8.

4. When 3,751 is divided by 31, what is the result?

5. Divide 1,911 by 21.

ANSWERS TO SELF-TEST 2.1

1. 1,428   2. 2,662   3. 51,612

4. 26,730   5. 15,120

ANSWERS TO SELF-TEST 2.2

1. 113   2. 3,936   3. 602

4. 121   5. 91

3 Focus on Multiplication

In this chapter, we'll make reference to Chapter 2, "Essential Arithmetic," pay a bit of attention to multiplication problems that demand greater time and attention, look at a few real-life uses of your multiplication skills, and consider when it's appropriate to use a calculator. We'll be doing some longer, more complex multiplication problems later in this chapter, but first, spend a little time checking your knowledge of multiplication facts. Having these in memory will make multiplication and estimation quicker and easier.

1 THE MULTIPLICATION TABLE

Table 3.1 goes from 1 × 1 to 10 × 10. Many tables go up to 12 × 12. But how many times do you need to recall 12 × 11?

How well do you know the table? There's only one way to find out. Fill it in. Multiply each number in the vertical column by each number in the horizontal row. A few are done to get you started.

Once you've finished, use Table 3.2 to check your work. Circle any wrong answers. The problems most commonly missed are 9 × 6 and 8 × 7.

Use Table 3.2 to check your answers.

Table 3.1   Blank Multiplication Table

Please fill in completely. If you're not too sure of yourself, do it in pencil first.										
	1	2	3	4	5	6	7	8	9	10
1	1	2								
2	2	4								
3			9	12						
4										
5										
6										
7										
8										
9										
10										

Table 3.2   Completed Multiplication Table

	1	2	3	4	5	6	7	8	9	10
1	1	2	3	4	5	6	7	8	9	10
2	2	4	6	8	10	12	14	16	18	20
3	3	6	9	12	15	18	21	24	27	30
4	4	8	12	16	20	24	28	32	36	40
5	5	10	15	20	25	30	35	40	45	50
6	6	12	18	24	30	36	42	48	54	60
7	7	14	21	28	35	42	49	56	63	70
8	8	16	24	32	40	48	56	64	72	80
9	9	18	27	36	45	54	63	72	81	90
10	10	20	30	40	50	60	70	80	90	100

If you got everything right, you can skip over to Frame 3. If you got just one or two of these wrong, then you'll need to go over them several times, until you're sure you know them.

What if you just don't know the multiplication table at all? Keep reading.

2   LEARNING THE MULTIPLICATION TABLE

Let's return to a concept discussed earlier: multiplication as addition. You can complete the number sequences below by adding the same number again and again. You might have heard it called skip counting. Fill in the missing numbers of these five series:

1	2	3	__	__	__	__	__	__	__
2	4	6	__	__	__	__	__	__	__
3	6	9	__	__	__	__	__	__	__
4	8	12	__	__	__	__	__	__	__
5	10	15	__	__	__	__	__	__	__

Now check your work:

1	2	3	4	5	6	7	8	9	10
2	4	6	8	10	12	14	16	18	20
3	6	9	12	15	18	21	24	27	30
4	8	12	16	20	24	28	32	36	40
5	10	15	20	25	30	35	40	45	50

If you've got those, you're halfway through the multiplication table. Try the sixes. Even if you need to count on your fingers, that's okay. People who count on their fingers are the original digital computers.

| 6 | 12 | 18 | — | — | — | — | — | — | — |

Sevens are next:

| 7 | 14 | 21 | — | — | — | — | — | — | — |

Now check your work:

| 6 | 12 | 18 | 24 | 30 | 36 | 42 | 48 | 54 | 60 |
| 7 | 14 | 21 | 18 | 35 | 42 | 49 | 56 | 63 | 70 |

How did you do? If you got them all right, go on to the next set. If not, continue working on those you missed.

You're ready to do eights:

| 8 | 16 | 24 | — | — | — | — | — | — | — |

Now try nines:

| 9 | 18 | 27 | — | — | — | — | — | — | — |

Finally, the easy one. Fill in the missing numbers for tens:

| 10 | 20 | 30 | — | — | — | — | — | — | — |

Now check your work:

8	16	24	32	40	48	56	64	72	80
9	18	27	36	45	54	63	72	81	90
10	20	30	40	50	60	70	80	90	100

Need more practice? Fill in Table 3.3 and check your work against Table 3.2.

If you're still getting some of these wrong, then try flash cards. Write the problem on one side of an index card and the answer on the other side. Practice them until you have memorized the correct answer to each of these problems.

Don't get discouraged. Once you've mastered these problems, you'll own them for the rest of your life. The only catch is that you'll have to keep using them.

As you build your memory of basic facts, you'll find things getting much easier. Numbers will seem a lot less intimidating, and you'll find yourself gaining confidence.

Table 3.3   Another Blank Multiplication Table

	1	2	3	4	5	6	7	8	9	10
1	1	2								
2	2	4								
3			9	12						
4										
5										
6										
7										
8										
9										
10										

We're now ready to apply what we've learned here by using the multiplication table in larger problems. Remember, if you don't use it, you'll lose it.

3   TESTING YOUR KNOWLEDGE

You may already know how to calculate more complex multiplication problems. There's only one way to find out—sink or swim. Please work out these three problems:

Problems:

$$\begin{array}{ccc}
89 & 195 & 7604 \\
\times\,57 & \times\,473 & \times\,3978
\end{array}$$

Solutions:

$$\begin{array}{r}
7604 \\
\times\,3978 \\
\hline
60832 \\
532280 \\
6843600 \\
22812000 \\
\hline
30248712
\end{array}$$

$$\begin{array}{r}
195 \\
\times\,473 \\
\hline
585 \\
13650 \\
78000 \\
\hline
92235
\end{array}$$

$$\begin{array}{r}
89 \\
\times\,57 \\
\hline
623 \\
4450 \\
\hline
5073
\end{array}$$

By the way, you may have learned to do multiplication like this without using those zeros to position each partial product on its line. That's fine if it's what you're used to. The zeros are just a reminder of why you shift over and a check that you have moved over the correct number of spaces.

Also did you notice that our answers had no commas? Certainly 30248712 looks better this way: 30,248,712. But we've left out the commas so our numbers would align vertically. To avoid mistakes, when you do long multiplication problems, make sure that your figures are aligned. And since you're not type-setting a book, be sure to put in all the commas.

If you got just the first problem right, then you're definitely in the right chapter. If all three problems were correct, then proceed to Frame 8. Don't worry, we won't be covering any new material before then.

If you answered two out of three problems correctly, keep reading for the next few frames, work out a few more problems, and then, if you feel you really know how to do long multiplication, you may want to skip to Frame 8.

4 ABOUT LONG MULTIPLICATION

Long multiplication repeats simple multiplication two or more times and finishes the work with addition of all the simple multiplications. Let's analyze the first problem. When we multiply 89 × 57, we're multiplying 89 × 7, then 89 × 50, which means 89 × 5 and put a zero at the end, and then adding the two products.

Let's walk through all the steps:

1. Begin with
$$\begin{array}{r} 89 \\ \times\,57 \end{array}$$

2. Multiply 7 × 9 = 63.

3. Write down the 3 and carry the 6. It represents 6 tens or sixty, so it gets carried over into the tens column.

$$\begin{array}{r} {}^{6} \\ 89 \\ \times\,57 \\ \hline 3 \end{array}$$

4. Multiply 7 × 8 = 56 and add on the carried 6 = 62.

5. Write down 62 to finish the first partial product.

$$\begin{array}{r} 89 \\ \times\,57 \\ \hline 623 \end{array}$$

6. Begin multiplying by 50 by placing a zero in the rightmost space of the next line, under the 3 of 623. Some people prefer to just leave this space blank and that's fine as long as you don't forget.

7. Multiply $5 \times 9 = 45$.

8. Write down the 5 and carry the 4.

$$
\begin{array}{r}
^4 \\
89 \\
\times\,57 \\
\hline
623 \\
50
\end{array}
$$

9. Multiply $5 \times 8 = 40$ and add the carried $4 = 44$.

10. Write down 44 to finish the second partial product.

$$
\begin{array}{r}
89 \\
\times\,57 \\
\hline
623 \\
4450 \\
\hline
5073
\end{array}
$$

11. Add $623 + 4450 = 5073$.

Notice how straight the columns are. One number right under another. You can avoid a lot of mistakes if you keep your columns straight.

5 PRACTICE EXERCISES WITH TWO DIGITS

For further practice, work out the next two problems and check the solutions:

Problems:

$$
\begin{array}{r}
64 \\
\times\,94
\end{array}
\qquad
\begin{array}{r}
59 \\
\times\,30
\end{array}
$$

Solutions:

$$
\begin{array}{r}
64 \\
\times\,94 \\
\hline
256 \\
5760 \\
\hline
6016
\end{array}
\qquad
\begin{array}{r}
59 \\
\times\,30 \\
\hline
00 \\
1770 \\
\hline
1770
\end{array}
$$

How did you do? If you got these two right, go on to Frame 6. If not, then let's spend some time going over both problems.

The first one is straightforward: $4 \times 4 = 16$, put down the 6 and carry the 1; $4 \times 6 = 24$; $24 + 1$ we carried $= 25$. So, we have 256 in the first row.

We begin the second row by multiplying 9 × 4 = 36. (If you wish, start the row by placing a zero under the 6 of 256, to remind you that you're really multiplying by 40. We must always place the appropriate number of zeros or indent by one additional place when we start a new row.) Place the 6 of 36 directly under the 5 of 256 and carry the 3. Next, we multiply 9 × 6 = 54 and 54 + 3 we carried = 57. This gives us 576 in the second row. Then we add.

Now we'll examine the second problem. How much is zero times any number? It's zero. We'll put two zeros in the first row. Remembering to indent, we multiply 3 × 9 = 27, placing the 7 in the second row and carrying the 2. Then 3 × 5 = 15 and 15 + 2 = 17, giving us 177 in the second row. The rest is addition. In this way, complex multiplication is broken down into a series of simple multiplication and addition steps.

Solve these problems:

Problems:

$$
\begin{array}{r}
80 \\
\times\,97 \\
\end{array}
\qquad
\begin{array}{r}
63 \\
\times\,50 \\
\end{array}
$$

Solutions:

$$
\begin{array}{r}
80 \\
\times\,97 \\
\hline
560 \\
7200 \\
\hline
7760 \\
\end{array}
\qquad
\begin{array}{r}
63 \\
\times\,50 \\
\hline
00 \\
3150 \\
\hline
3150 \\
\end{array}
$$

If you got this right, go on to Frame 6; if not, check to be sure you know the multiplication table. If you don't, you'll need to go back to the beginning of Frame 2. If, on the other hand, you do know the table, then begin rereading this Frame.

6   MULTIPLYING THREE DIGITS

If you've successfully multiplied two-digit numbers, you may be able to anticipate how to multiply larger numbers. Try to solve these two problems:

Problems:

$$
\begin{array}{r}
672 \\
\times\,305 \\
\end{array}
\qquad
\begin{array}{r}
580 \\
\times\,941 \\
\end{array}
$$

Solutions:

$$
\begin{array}{r}
672 \\
\times\,305 \\
\hline
3360 \\
0000 \\
201600 \\
\hline
204960
\end{array}
\qquad
\begin{array}{r}
580 \\
\times\,941 \\
\hline
580 \\
2320 \\
5220 \\
\hline
545780
\end{array}
$$

If you got both problems right, go on to Frame 7; if not, keep reading.

Let's go over the first problem. The first row is straightforward: $5 \times 2 = 10$; write down 0 and carry 1; $5 \times 7 = 35$ and $35 + 1 = 36$; write down 6 and carry the 3; $5 \times 6 = 30$ and $30 + 3 = 33$.

The next row is all zeros. The third row: $3 \times 2 = 6$; $3 \times 7 = 21$; write down 1 and carry 2; $3 \times 6 = 18$ and $18 + 2$ we carried = 20. Then add.

Here's an alternate way of doing this problem:

$$
\begin{array}{r}
672 \\
\times\,305 \\
\hline
3360 \\
201600 \\
\hline
204960
\end{array}
$$

When you multiply by zero, you get zero, and adding zero doesn't change the total, so you can skip over the multiplying by zero. Just be careful to add the right number of zeros or indent the correct number of places for the next line. When you're at ease with numbers, you can do it this way. But if you're not yet that much at ease, continue writing out all the zeros.

The second problem is a little easier. The first row is easily obtained. $1 \times 580 = 580$. In the second row, $4 \times 0 = 0$; $4 \times 8 = 32$; write down the 2 and carry the 3; $4 \times 5 = 20$ and $20 + 3 = 23$. In the third row, $9 \times 0 = 0$; $9 \times 8 = 72$; write down the 2 and carry the 7; $9 \times 5 = 45$ and $45 + 7 = 52$. Then add.

Here's another set of problems:

Problems:

$$
\begin{array}{r}
905 \\
\times\,658
\end{array}
\qquad
\begin{array}{r}
837 \\
\times\,405
\end{array}
$$

Solutions:

(The zeros on the end of the partial products aren't shown on this set, but if you like, you can fill them in.)

$$
\begin{array}{r}
905 \\
\times\,658 \\
\hline
7240 \\
4525 \\
5430 \\
\hline
595490
\end{array}
\qquad
\begin{array}{r}
837 \\
\times\,405 \\
\hline
4185 \\
000 \\
3348 \\
\hline
338985
\end{array}
$$

Try to solve this pair of problems:

Problems:

$$
\begin{array}{r}
416 \\
\times\,807 \\
\end{array}
\qquad
\begin{array}{r}
590 \\
\times\,472 \\
\end{array}
$$

Solutions:

$$
\begin{array}{r}
416 \\
\times\,807 \\
\hline
2912 \\
000 \\
3328 \\
\hline
335712 \\
\end{array}
\qquad
\begin{array}{r}
590 \\
\times\,472 \\
\hline
1180 \\
4130 \\
2360 \\
\hline
278480 \\
\end{array}
$$

7 FOUR-DIGIT MULTIPLICATION

Now it's on to four-digit multiplication. The process is the same, you'll just have more rows. To be honest, four-digit numbers times four-digit numbers is when many people start looking for their calculators, and that's fine. Let's work out these two problems, but before you do, try to make an estimate of the answer. It will help you check your answer, whether you do it by hand or use a calculator.

Problems:

$$
\begin{array}{r}
5076 \\
\times\,3245 \\
\end{array}
\qquad
\begin{array}{r}
6975 \\
\times\,4056 \\
\end{array}
$$

Solutions:

For the first problem, estimate $5,000 \times 3,000$. $5 \times 3 = 15$ and we need 6 zeros, so our answer should be around 15,000,000. (Need help with this? See Chapter 5, "Mental Math.") We rounded both numbers down, so our estimate is probably a little low.

$$
\begin{array}{r}
5076 \\
\times\,3245 \\
\hline
25380 \\
20304 \\
10152 \\
15228 \\
\hline
16,471,620 \\
\end{array}
$$

For the second problem, round 6,975 up to 7,000 and 4,056 down to 4,000. 7,000 × 4,000 = 28,000,000.

$$
\begin{array}{r}
6975 \\
\times\ 4056 \\
\hline
41850 \\
34875 \\
0000 \\
27900 \\
\hline
28{,}290{,}600
\end{array}
$$

If you got these right, you can skip to Frame 8 of this chapter. If you got either (or both) of these wrong, here's your chance to redeem yourself:

Problems:
$$
\begin{array}{r}
1904 \\
\times\ 6578 \\
\hline
\end{array}
\qquad
\begin{array}{r}
4714 \\
\times\ 5062 \\
\hline
\end{array}
$$

Solutions:

Estimate the first problem as 2,000 × 6,000 and expect an answer around 12,000,000.

$$
\begin{array}{r}
1904 \\
\times\ 6578 \\
\hline
15232 \\
13328 \\
9520 \\
11424 \\
\hline
12{,}524{,}512
\end{array}
$$

For the second problem, estimate 5,000 × 5,000 = 25,000.

$$
\begin{array}{r}
4714 \\
\times\ 5062 \\
\hline
9428 \\
28284 \\
0000 \\
23570 \\
\hline
23{,}862{,}268
\end{array}
$$

Okay. Now it's time to see how well you can do multiplication with several digits. Do the following problems and check your work. If you were correct on all five, go on to Frame 8. If not, go back and redo Frames 4–7 and then retake Self-Test 1.

SELF-TEST 3.1

1.   95
      × 32

2.   781
     × 450

3.   446
     × 108

4.   7509
     × 5124

5.   9087
     × 1703

8 | WORD PROBLEMS WITH MULTIPLICATION

Everyone eventually can multiply one number by another. That's straightforward. But if the same problem is stated in words rather than numbers, it becomes much more confusing to many people. To get over this fear and confusion, let's tackle some word problems.

Before we start, remember that hidden in each of these word problems is one ordinary multiplication problem. All you'll need to do is set up each multiplication problem and solve it.

Problem 1:

A department store is having a sale on French perfume at $90 a bottle. How much would you have to pay for 12 bottles?

Solution:

$90 × 12

$$
\begin{array}{r}
\$90 \\
\times\,12 \\
\hline
180 \\
900 \\
\hline
\$1080
\end{array}
$$

Problem 2:

An appliance store sold 234 VCRs at $235 each. How much is the store's total sales of VCRs?

Solution:

$235 × 234

$$
\begin{array}{r}
\$235 \\
\times\,234 \\
\hline
940 \\
7050 \\
47000 \\
\hline
\$54990
\end{array}
$$

Have you gotten the hang of it? Here's one last one.

Problem 3:

If light travels at the rate of 186,000 miles per second, how far does it travel in two minutes? (Remember: there are 60 seconds in 1 minute, or 120 seconds in 2 minutes.)

Solution:

186,000 miles × 120

$$
\begin{array}{r}
186000 \\
\times\,120 \\
\hline
3\,720\,000 \\
18\,600\,000 \\
\hline
22{,}320{,}000
\end{array}
$$

Now we'll see how well *you* do at translating words into numbers. You'll know after you've taken Self–Test 3.2.

SELF-TEST 3.2

1. A dealer sold 75 used cars at $10,995 each. What were her total sales?

2. If a train travels at an average speed of 70 miles per hour, how far has the train traveled in 14 hours?

3. There are 5,280 feet in a mile. If you walked 18 miles, how many feet did you walk?

ANSWERS TO SELF-TEST 3.1

$$
\begin{array}{r}
1. \quad 95 \\
\times\,32 \\
\hline
190 \\
2\,850 \\
\hline
3{,}040
\end{array}
\qquad
\begin{array}{r}
2. \quad 781 \\
\times\,450 \\
\hline
39\,050 \\
312\,400 \\
\hline
351{,}450
\end{array}
\qquad
\begin{array}{r}
3. \quad 446 \\
\times\,108 \\
\hline
3\,568 \\
44\,600 \\
\hline
48{,}168
\end{array}
$$

4.	7509	5.	9087
	× 5124		× 1703
	30 036		27 261
	150 180		6 360 900
	750 900		9 087 000
	37 545 000		15,475,161
	38,476,116		

ANSWERS TO SELF-TEST 3.2

1.	$10995	2.	70 miles	3.	5280 feet
	× 75		× 14		× 18
	54 975		280		42 240
	769 65		70		52 80
	$824,625		980 miles		95,040 feet

How did you do? If you got these—that's great. If you didn't, you're in good company. Most people have some trouble with word problems. We'll work more on word problems in later chapters.

4 Focus on Division

In the last chapter, we paid attention to more complicated multiplication problems; now it's time for more complicated division problems. Again, we'll be making reference back to Chapter 2, "Essential Arithmetic," investigating some applications of division in daily life, and considering when we want to choose a calculator to help with the work.

Division is the reverse of multiplication. For instance, $2 \times 5 = 10$; and $10 \div 5 = 2$. If you can multiply, you can divide. If multiplication is thought of as repeated addition, division can be thought of as repeated subtraction. How many times can you subtract 7 from 83? More than once, certainly. More than 10 times? Ten sevens would be 70, so yes, you can subtract 7 from 83 more than 10 times. $83 - 70 = 13$ so you can take away one more 7, but then there will only be 6 left. You subtracted 7 (the divisor) from 83 (the dividend) 11 times (the quotient) and had 6 left over (the remainder).

1 SHORT DIVISION

We'll deal with short division first, and we'll get to long division in the second part of this chapter. What's the boundary line between the long and the short? If the number we divide by is a single digit, then we're doing short division; if it's more than one digit, then we're doing long division. With experience, you may find you can do short division for larger divisors, but we'll focus on one-digit divisors for now.

Here is a pair of short division problems.

Problems:

$$9\overline{)360} \quad 5\overline{)365}$$

Solutions:

$$9\overline{)360}^{\,40} \quad 5\overline{)36^15}^{\,73}$$

If you got both of those right, go on to the next problem set; if not, read on.

Let's go over the first problem. We know that 9 won't go into 3, because 3 is smaller than 9, but it will go into 36. How many times? More than once? More than two times? More than three times? In fact, 9 goes into 36 exactly 4 times. Remember your multiplication table?

And next, how many times does 9 go into 0? Zero times, so the answer to our problem is 40. Let's check it out. How much is 9×40? It's 360.

In the second division problem, how many times does 5 go into 3? No times. How many times does 5 go into 36? More than 5? Yes! More than 6? Yes! More than 7? Yes—a little more than 7.

Here, what we do is write 7 as part of our answer and carry 1 because $5 \times 7 = 35$ but we started with 36. We make the final 5 in 365 into 15. Since 5 goes into 15 exactly 3 times, we have our answer: 73. This answer is verified by the multiplication problem $73 \times 5 = 365$.

The check is an important part of division, because by multiplying our answer times the number by which we divided (the divisor), and adding on any remainder we had, we should end up with the number we divided (the dividend). If not, we must go back and find our mistake.

Both problems in that set worked out nicely. (Some would say they "came out even" but the word "even" has a specific meaning in math that doesn't really fit here.) The divisor, the one-digit number, "went in" or could be subtracted, a certain number of times exactly, with nothing left over. There was no remainder.

What do we do when there is a remainder? That depends on the question you're trying to answer. You might just say "with 6 left over." You might talk about the divisor going in a fraction of a time. You might show the decimal point at the end of the number, add some zeros, and keep dividing.

If you bake 39 cookies and pack them up to give to friends with 4 cookies in each pack, how many packs of cookies can you make? 39 cookies ÷ 4 per pack = 9 packs with 3 cookies left over (for you). You could also say you have 9 packs and a part or fraction of another. The fraction is the number left over (3) over the number in a pack (4). You have $9\frac{3}{4}$ packs. But if you were trying to share 39 dollars among 4 people, you wouldn't give everyone $9 and just throw the remaining $3 away. You'd write $39 in dollars and cents form as 39.00 and keep dividing.

$$4\overline{)\,39.^30^20}\overset{\textstyle 9.75}{}$$

Each of the 4 people get $9.75.

When you're doing a division with nothing to tell you which way to deal with the remainder, you can choose your favorite. We'll show answers in different forms for different problems.

Here's the next set.

Problems:

$$8{\overline{)150}} \qquad 3{\overline{)191}}$$

Solutions:

$$
\begin{array}{r}
18.75 \\
\hline
8{\overline{)15^70.^60^40}}
\end{array}
\qquad
\begin{array}{r}
63\frac{2}{3} \\
\hline
3{\overline{)19^11}}
\end{array}
$$

How did you do? If you got these right, go on to the next problem set; if you didn't, then read the explanation here.

The first problem worked out once we put in a decimal point right after 150 and added a couple of zeros. What gives us the right to do this? Mathematically, every whole number has an unseen decimal point immediately to its right, followed by any number of zeros. For example, 8 may be written 8.0, or 8.00, or 8.000. The reason we can place a decimal point after a number and add zeros is because doing so doesn't alter the value of that number.

In the first problem, we added a decimal point after 150 and added, or annexed, two zeros. By continuing the division, we came up with an answer of 18.75. We need to be very careful about placing our decimal immediately to the right of a number before we add zeros. How many zeros do you need to add? Usually just one or two, but you're hoping to get to a place where there is no remainder.

Many division problems, even if you add the decimal point and zeros, will not get to a point where there is no remainder. Instead, you'll see the same remainder or series of remainders, repeating. This is true in the second problem. When we divide 3 into 191, we get 63, with two left over. No matter how far we carry our division, we'll keep coming up with two left over. You can choose to place the remainder over the divisor to make a fraction as we did in the second problem or you can stop dividing at some point and round the answer to a convenient number of decimal places, usually one or two. If we were to round the answer to the second problem to two decimals, we'd get 63.67. If we rounded to one decimal, we'd have 63.7.

Rounding

Suppose you have $16.25 in your pocket. How much do you have to the nearest dollar? Is what you have closer to $16 or to $17? You have about $16. If you earn $28,603, how much do you earn to the nearest thousand? Is your income closer to $28,000 or to $29,000? $28,500 would be halfway and you're past that so you earn approximately $29,000.

If you weigh 117.476 pounds, how much do you weigh to the nearest pound? Is your weight closer to 117 pounds or past 117.5 and so closer to 118? The answer is 117 pounds. What is your weight to the nearest tenth of a pound? Is it closer to 117.4 or 117.5? The 7 that follows the 4 in 117.476 is the signal that you're closer to 117.5. And your weight to the nearest hundredth of a pound is 117.48.

The digit that follows the one you want to round tells you whether to round up or down. If the next digit is 4 or less, drop the following digits but don't change the digit in the place you're rounding to. This is rounding down. If the following digit is 5 or greater, drop the following digits but increase the digit in the place you're rounding to by 1. That's rounding up. When we round a number ending in 5, the general practice is to round up (from 115.25 to 115.3) rather than down (from 115.25 to 115.2). However, if you are doing a lot of rounding and you can remember, the soundest practice is to alternate, by rounding upward in the first instance, downward in the second, and so forth.

Let's do another set of problems.

Problems:

$$9\overline{)179} \qquad 4\overline{)3142}$$

Solutions:

(For problems that never get to a point where the remainder is zero, we'll show different ways to format the answer.)

$$\frac{19\frac{8}{9}}{9\overline{)17^89}} \quad \text{or} \quad \frac{1\,9.\,8\,8\,8}{9\overline{)17^89.^80^80^80}} \qquad \text{Round to 19.89} \qquad \frac{7\,8\,5.\,5}{4\overline{)31^34^22.^20}}$$

If you came up with the right answers to these last two problems, go on to Frame 2; if not, try some more problems.

Problems:

$$7\overline{)473} \qquad 6\overline{)9613}$$

Solutions:

$$\frac{67\frac{4}{7}}{7\overline{)47^53}} \quad \text{or} \quad \frac{67.\,571}{7\overline{)47^53.^40^50^10}} \qquad \text{Round to 67.57}$$

$$\frac{1602\frac{1}{6}}{6\overline{)9^361^13}} \quad \text{or} \quad \frac{1602.\,166}{6\overline{)9^3613.^10^40^40}} \qquad \text{Round to 1602.17}$$

How did you do? Here's one last set.

Problems:

$$8\overline{)9077} \qquad 5\overline{)8154}$$

Solutions:

$$\begin{array}{c} 1\ 1\ 3\ 4.6\ 2\ 5 \\ 8\overline{)9^10^27^37.^50^20^40} \end{array} \qquad \begin{array}{c} 1\ 6\ 30.\ 8 \\ 5\overline{)8^31^154.^40} \end{array}$$

At this point you'll need to make a judgment. If you're having a tough time, the best thing to do would be to go back to the beginning of this chapter and start again from scratch. You won't be able to do the long division of the next section until you're good at short division.

One of the things we've been stressing is that mathematics is a set of tools that build on one another. You need to know the multiplication table to multiply. You can't divide until you learn how to multiply. And you can't do long division until you learn short division.

SELF-TEST 4.1

1. $6\overline{)450}$ 2. $7\overline{)497}$ 3. $4\overline{)183}$

4. $5\overline{)286}$ 5. $9\overline{)790}$

2   LONG DIVISION

Long division is carried out in five steps: (1) divide; (2) multiply; (3) subtract; (4) compare; and (5) bring down. Remembering the steps is easier if you have a silly sentence whose words begin with the letters DMSCB, the first letter of each step. You can try Don't Miss Susie's Chocolate Brownies or Don't Make Silly Computation Blunders, or Do Monkeys Sleep Completely Bare? Or make up one of you own.

The process of long division is identical to short division, but it involves a lot more calculation and so we write down a lot of the thinking that, during short division, we were able to do in our heads. It's important to have memorized the multiplication table. Estimation is key to getting started and you'll need to know your tables for that. Because these problems are relatively long, we'll take them one at a time and round the answers to the nearest tenth when necessary.

Problem 1:

$$43 \overline{\smash{)}\,795}$$

Solution:

Round 18.48 to 18.5.

$$
\begin{array}{r}
18.48 \\
43 \overline{\smash{)}\,795.00} \\
{\scriptstyle x\ xx} \\
-43 \\
\hline
365 \\
-344 \\
\hline
210 \\
-172 \\
\hline
380 \\
-344 \\
\hline
36
\end{array}
$$

First, because we anticipate this will not come out with a zero remainder, we place the decimal point after the 5 and add two zeros. They won't do any harm if it turns out we don't need them.

Next, estimate that 43 will go into 79 only one time, because 2 × 43 is 86, which is more than 79. Multiply 1 × 43 and place 43 under 79. Subtract 43 from 79, leaving 36, and compare 36 to the divisor of 43. The result of the subtraction should be less than the divisor, and it is, so continue.

Bring down the 5 (keep track of what's been brought down by placing an X underneath, especially if there are several of the same digit).

How many times does 43 go into 365? Estimate with 40 and 360. That would be 9. Multiply 9 × 43 and you find that's a little too big (387) so use 8 instead of 9. Say 43 goes into 365 eight times. Now 8 × 43 = 344. We subtract 344 from 365 and get 21, which is less than the divisor. Bring down the first zero and estimate 43 goes into 210 four times. (200 ÷ 40 would be 5, but again 5 × 43 is a little too big.) Since 4 × 43 = 172, subtract 172 from 210 and get 38. Since 38 is less than 43, continue.

Bring down the second zero. How many times does 43 go into 380? Remember from earlier 8 × 43 = 344 so put 8 in the quotient. (Doing your checks in the margin often saves you from doing the same work more than once.) That's far enough, since the instructions were to round to one decimal digit, we round off 18.48 to 18.5.

Problem 2:

$$57 \overline{\smash{)}\,2075}$$

Solution:

$$
\begin{array}{r}
36.4 \\
57\overline{)2075.0} \\
\scriptstyle{x\ x} \\
-171 \\
\hline
365 \\
-342 \\
\hline
230 \\
-228 \\
\hline
2 \\
\end{array}
$$

How many times does 57 go into 207? It goes in three times, so write 3 over the 7, multiply 3 × 57 and get 171. Subtract 171 from 207 and get 36, which is less than the divisor of 57. Bring down the next digit, 5, for 365.

How many times does 57 go into 365? It goes in six times, so multiply 57 × 6 and get 342. Subtract 342 from 365 to get 23, which is less than 57, so bring down the zero to make 230.

How many times does 57 go into 230? Remember 4 × 57 = 228, so place the 4 up top and subtract 228 from 230 to get 2.

We'll add one more wrinkle and that will be it for long division. We'll divide by a three–digit number. Use your estimating skills and try this one.

Problem 3:

$$614\overline{)1437}$$

Solution:

Estimate by thinking about how many times 600 would go into 1,400. It would go twice, but three times would be 1,800 so that's too big. We'll expect a quotient a little more than 2. If you want to get a better estimate, remember that $600 \times 2 = 1,200$ which means there would be about 200 left over. Our estimate would be about $2\frac{200}{600} = 2\frac{2}{6} = 2\frac{1}{3}$. The division gives us 2.34. Round 2.34 to 2.3.

$$
\begin{array}{r}
2.34 \\
614\overline{)1437.00} \\
\scriptstyle{x\ x} \\
-1228 \\
\hline
2090 \\
-1842 \\
\hline
2480 \\
-2456 \\
\hline
24 \\
\end{array}
$$

And now, one last problem.

Problem 4:

$$591\overline{)83902}$$

Solution:

Estimate $84{,}000 \div 600$. That's a little more difficult, but we can ignore two zeros from each number and think about $840 \div 6$. That will be over 100, for sure, but let's try to get closer. $840 \div 6 = 140$, so expect an answer around 140. After the division, round 141.96 to $141.^10 = 142.0$.

$$
\begin{array}{r}
141.96 \\
591\overline{)83902.00} \\
\text{xx xx} \\
-591 \\
\hline
2480 \\
-2364 \\
\hline
1162 \\
-591 \\
\hline
5710 \\
-5319 \\
\hline
3910 \\
-3546 \\
\hline
364
\end{array}
$$

Wasn't that a whole lot of fun? Now see what you can do with the problems in Self–Test 4.2.

SELF-TEST 4.2

1. $19\overline{)306}$ 2. $67\overline{)541}$ 3. $239\overline{)1975}$

4. $641\overline{)84330}$ 5. $785\overline{)75411}$

3 WORD PROBLEMS WITH DIVISION

Here they come again. I hope these won't be traumatic. Just remember: make each word problem into a simple numerical problem. One advantage of a word problem is that the situation described in the problem will likely tell you whether to express a remainder as a fraction or a decimal or just "left over." And it may help you decide what place to round a decimal answer.

Problem 1:

On a religious holiday, one-eighth of the 5,000-member student body of a high school stayed home to observe. How many students were absent and how many came to school?

Solution:

To find one-eighth of a number, divide by 8. 5,000 ÷ 8. Subtract that answer from 5,000 to find out how many came to school.

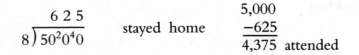

$$\begin{array}{r} 6\,2\,5 \\ 8\,\overline{)\,50^20^40} \end{array}$$ stayed home $$\begin{array}{r} 5{,}000 \\ -625 \\ \hline 4{,}375 \end{array}$$ attended

Problem 2:

In a school yard there are 264 children. How many soccer teams of 11 players each can be formed?

Solution:

264 ÷ 11

$$\begin{array}{r} 24 \\ 11\,\overline{)\,26^44} \end{array}$$

24 teams can be formed and no student is left out.

Problem 3:

Melissa Jones earned $40,000 last year. How much did she earn per week? (There are 52 weeks in a year.)

Solution:

$40,000 ÷ 52

$$\begin{array}{r} \$769.23 \\ 52\,\overline{)\,\$40000.00} \\ \text{xx xx} \\ -364 \\ \hline 360 \\ -312 \\ \hline 480 \\ -468 \\ \hline 120 \\ -104 \\ \hline 160 \\ -156 \end{array}$$

That last problem is a place where estimation might help, and might even be enough, depending on why you're asking the question. Round 52 down to 50 and divide $40,000 ÷ 50, which is equivalent to 4,000 ÷ 5 = 800. She earns a bit less than $800 a week.

Now try your luck at Self–Test 4.3.

SELF-TEST 4.3

1. If a rope is 144 inches long, how many yards long is that rope? (A yard is 3 feet or 36 inches.)

2. A government agency had a payroll of $7,500,000 for 864 workers who all earned exactly the same annual pay. How much did each worker earn?

3. Spuds, the neighbor's dog, chews up 16,316 bones a year. How many bones does he go through in a day?

ANSWERS TO SELF-TEST 4.1

1.
$$6\overline{)45^30} = 75$$

2.
$$7\overline{)497} = 71$$

3.
$$4\overline{)18^23.^30^20} = 45.75$$

4.
$$5\overline{)28^36.^10} = 57.2$$

5.
$$9\overline{)79^70} = 87\tfrac{7}{9}$$ or $$9\overline{)79^70.^70^70} = 87.77$$ Round to 87.8

ANSWERS TO SELF-TEST 4.2

1. If you kept bringing down zeros and dividing, the next digit would be zero, so you'd round to 16.1.

$$
\begin{array}{r}
16.1 \\
19\overline{)306.0} \\
{}_{\text{x x}} \\
-19 \\
\hline
116 \\
-114 \\
\hline
20 \\
-19 \\
\hline
1
\end{array}
$$

2. Round 8.07 to 8.1.

```
              8.07
        67) 541.00
                 x x
           −536
            500
           −469
             31
```

3. Round 8.26 to 8.3.

```
               8.26
        239) 1975.00
                  x x
            −1912
              630
             −478
             1520
            −1434
               86
```

4. Round 131.56 to 131.6.

```
               131.56
        641) 84330.00
                  x x  x x
            −641
            2023
           −1923
            1000
            −641
            3590
           −3205
            3850
           −3846
               4
```

5. Round 96.06 to 96.1.

```
               96.06
        785) 75411.00
                   x  x x
            −7065
            4761
           −4710
            5100
           −4710
             390
```

1.

$$
\begin{array}{r}
4 \text{ yards} \\
36\overline{)144} \\
-144 \\
\hline
\end{array}
$$

2. Round $8,680.555 to $8,680.56.

$$
\begin{array}{r}
\$8680.555 \\
864\overline{)\$7500000.000} \\
{\scriptstyle x x x\ x x x} \\
-6912 \\
\hline
5880 \\
-5184 \\
\hline
6960 \\
-6912 \\
\hline
4800 \\
-4320 \\
\hline
4800 \\
-4320 \\
\hline
4800 \\
-4320 \\
\hline
480
\end{array}
$$

3. Would a dog eat part of a bone? This answer of 44.7 is closer to 45 bones.

$$
\begin{array}{r}
44.7 \\
365\overline{)16316.0} \\
{\scriptstyle x\ x} \\
-1460 \\
\hline
1716 \\
-1460 \\
\hline
2560 \\
-2555 \\
\hline
5
\end{array}
$$

If you're starting to feel that a calculator would make some of these much easier, you're probably right, but it's wise to know how to do the tough problems so you can make intelligent choices about when to use a calculator.

5 Mental Math

Did you know that you could multiply many numbers by 10 just by adding a zero? And that you could divide many numbers by 100 by shifting a decimal point two places to the left? Those are just two of the shortcuts that can help you do calculations "in your head," without needing paper and pencil or a calculator. In this chapter, we'll look at some of the tips and tricks for mental math. You may not need all of them, but you may find some helpful, and they're also good for amazing your friends. We'll be making reference back to Chapter 2, "Essential Arithmetic," so review that if necessary.

1 FAST MULTIPLICATION

One of the laws of arithmetic is that any number multiplied by 1 *is* that number. The next time someone asks you to multiply 10 by 1, you'll know the answer is 10. And 3 × 1 is 3. Multiplying by 1 doesn't change anything.

What about multiplying by 10? How much is 16 × 10? If you were to do this with paper and pencil, 0 times 16 would just be 0, and then you'd shift over one place and multiply 1 times 16, which is just 16. When you add it up, it's 160. What we've done, in effect, is move the decimal one place to the right (from 16.0 to 160.0). We don't usually show the decimal point for a whole number like 16, so it just looks like adding a zero.

Problems:
Multiply these numbers by 10:

(a) 4

(b) 37

(c) 596

Solutions:

(a) 4 × 10 = 40 (4.0 to 40.0)

(b) 37 × 10 = 370 (37.0 to 370.0)

(c) 596 × 10 = 5,960 (596.0 to 5,960.0)

Multiplying by 10 looks like adding a zero to the number being multiplied. That observation is fine as long as we're dealing with whole numbers. But when we deal with decimals, we've got to worry about the decimal point.

How much is 10 × 0.9? This is where you need to realize that you're really moving the decimal point, so move it: 10 × 0.9 is 09.0. The answer is 9.0, or simply 9. How did we get that? We moved the decimal one place to the right.

But what if we had merely added a zero? Then we would have gotten .90, which, mathematically, is equal to .9 because $.90 = \frac{90}{100} = \frac{9}{10} = .9$ Adding a zero after the digits to the right of the decimal point doesn't make the number ten times larger. It's just an alternate name for the same number. When we're multiplying a decimal by 10, we have to make sure to move the decimal one place to the right. Let's try some problems with decimals.

Problems:

Multiply these numbers by 10:

(a) 6.3

(b) 0.4

(c) 0.02

Solutions:

(a) 63

(b) 4

(c) 0.2

Our next step is to multiply by 100. To do that we add two zeros to whole numbers or move the decimal point two places to the right in decimals.

Problems:

Multiply each of these numbers by 100:

(a) 14

(b) 1

(c) 0.7

(d) 0.05

Solutions:

(a) 1,400

(b) 100

(c) 70

(d) 5

The rule you've learned for multiplying by 10 and by 100 will also work, with appropriate adjustment, for multiplying by 1,000 or 10,000 or any number that's a 1 followed by zeros. $4{,}933 \times 1{,}000 = 4{,}933{,}000$ and $5.82 \times 10{,}000 = 58{,}200$.

Problems:

Multiply each of these as shown:

(a) $5 \times 1{,}000$

(b) $23 \times 1{,}000{,}000$

(c) $0.8 \times 10{,}000$

(d) $0.27 \times 10{,}000{,}000$

Solutions:

(a) 5,000

(b) 23,000,000

(c) 8,000

(d) 2,700,000

Let's see what you've learned so far. See if you can complete Self-Test 5.1 without writing anything down.

SELF-TEST 5.1

1. Multiply each of these numbers by 10:

    (a) 0.09   (b) 0.7   (c) 86

    (d) 5.6   (e) 102

2. Multiply each of these numbers by 100:

(a) 4   (b) 0.02   (c) 17

(d) 0.13   (e) 0.008

3. Complete each of these multiplications:

(a) 597 × 10,000   (b) 598.16 × 1,000,000   (c) 3,812 × 10,000,000

(d) 11.729 × 100,000   (e) 5.1 × 1,000,000,000

2 FAST DIVISION

Fast division is the exact reverse of fast multiplication. Instead of moving the decimal place to the right, we move it to the left. If you divide 43 by 10, 10 goes in 4 times with 3 left over. You could write that as $4\frac{3}{10}$ or 4.3. The decimal that was after the 3 in 43 moved to the left and now is between 4 and 3. And in the case of whole numbers ending with zeros, we can just remove zeros. How many times does 10 go into 1,000? You could do the work to find out it goes in 100 times or just remove one zero from the end of 1,000.

Are you ready for some problems? Please do these in your head.

Problems:

Divide each of the following numbers by 10:

(a) 800

(b) 16

(c) 0.3

Solutions:

(a) 80

(b) 1.6

(c) 0.03

In each case what we did was move the decimal one place to the left. In the first case, 800.0 became 80.0; in the second, 16 became 1.6; and in the third, 0.3 became 0.03.

Now we'll divide by 100. All you'll do here is move the decimal *two* places to the left. You can write in zeros if you don't have enough places.

Problems:

Please divide each of these numbers by 100:

(a) 0.6

(b) 100

(c) 9

Solutions:

(a) 0.006

(b) 1

(c) 0.09

Just as we saw with multiplication, this trick will work for any divisor that is a 1 followed by zeros, and the number of zeros will tell you how many places to move the decimal point.

Problems:

Please divide each of these:

(a) 475 ÷ 1,000

(b) 287.5 ÷ 100,000

(c) 0.04 ÷ 1,000,000

Solutions:

(a) 3 places left: 475.0 = 0.475

(b) 5 places left: 00287.5 = 0.002875

(c) 6 places left: 000000.04 = 0.00000004

I'll bet you're finding these pretty easy. Try Self-Test 5.2.

SELF-TEST 5.2

1. Divide each of these numbers by 10:

(a) 0.01   (b) 4   (c) 900

(d) 0.71   (e) 0.3

2. Divide each of these numbers by 100:

    (a) 80   (b) 0.14   (c) 3.7

    (d) 916   (e) 0.05

3. Do each division mentally:

    (a) 9,432 ÷ 1,000   (b) 18.1 ÷ 10,000   (c) 594.71 ÷ 1,000,000

    (d) 6,542,038 ÷ 100,000 (e) 17.8635 ÷ 100,000,000

3 | HOW ARE YOU DOING SO FAR?

If the pace we've been setting is too fast for you, then you would do well to turn back to the beginning of Chapter 2. Mathematics is a lot like building blocks. If you don't get the foundation down solidly, your building may come tumbling down. Even if this sets you back a bit, in the long run, you'll do much better.

If the pace is too slow for you, then go ahead and try the self-tests. If you do well, then you may skip the explanations and go on to the next chapter.

4 | MENTAL MATH TRICKS

The tricks for fast multiplication and fast division that you learned in Frame 1 and Frame 2 can be adapted to work in other situations than just multiplying by a number that's a 1 followed by zeros. To make them easier to describe, let's say those numbers are called powers of 10. (There's more on powers—of ten or anything else—in Chapter 15, Solving Simple Equations, but for now, we just want the name.)

Break It Down

It's good to have quick ways to work with powers of 10, but the need to multiply doesn't always come with numbers quite that round. If you need to multiply 51 by 102, multiplying 51 by 100 will give you a good estimate, but the exact value is easy enough to find if you break the job into two easier problems: 51×100 and 51×2. The first just requires adding zeros: $51 \times 100 = 5,100$. The second bit isn't all that hard, and if you wanted to, you could break it down into 50×2 and 1×2. Either way, $51 \times 2 = 102$. Add that 102 to the 5,100 and you've got $5,100 + 102 = 5,202$.

Problems:

Multiply each of the following by breaking the problem into smaller problems:

(a) $24 \times 1{,}010$

(b) 38×12

(c) 105×16

Solutions:

(a) $24 \times 1{,}010 = 24 \times 1{,}000 + 24 \times 10 = 24{,}000 + 240 = 24{,}240$

(b) $38 \times 12 = 38 \times 10 + 38 \times 2 = 380 + 30 \times 2 + 8 \times 2 = 380 + 60 + 16 = 440 + 16 = 456$

(c) $105 \times 16 = 100 \times 16 + 5 \times 16 = 1{,}600 + 5 \times 10 + 5 \times 6 = 1{,}600 + 50 + 30 = 1{,}680$

You may have seen a different way to break those down than the one we used. For example, instead of thinking of 105×16 as $100 \times 16 + 5 \times 16$, you might have said $105 \times 10 + 105 \times 6$. Did you get the same answer at the end? Then you're fine. Use whichever breakdown makes sense to you.

Let's add a little more multiplication power to your mental math arsenal. We've talked about how to multiply or divide by a power of ten. Now we'll see how to use that idea to multiply or divide by numbers like 200 or 5,000.

Multiples of Powers of 10

You know that $84 \times 100 = 8{,}400$, but what if you needed to multiply 84×200 or 84×500? This is another place where breaking the problem down into more manageable pieces can be a great help. You could think of 84×200 as $84 \times 100 + 84 \times 100$, but that gets cumbersome if you're multiplying by 500 or 900 instead of 200. (If you like it, it will work.) Let's try a slightly different way to think about it: $84 \times 2 \times 100$. Do 84×2 first (you can break it into $80 \times 2 + 4 \times 2$ if you want to. $84 \times 2 \times 100 = 168 \times 100 = 16{,}800$.

Can you multiply 35×222? Break it down first: $35 \times 200 + 35 \times 20 + 35 \times 2$. If you can see that 35×2 is 70, you've done most of the work.

$$35 \times 200 + 35 \times 20 + 35 \times 2 = 35 \times 2 \times 100 + 35 \times 2 \times 10 + 35 \times 2$$
$$= 70 \times 100 + 70 \times 10 + 70$$
$$= 7{,}000 + 700 + 70$$
$$= 7{,}770$$

Problems:

Multiply each of the following by breaking them down into simpler problems:

(a) 25×111

(b) 22×233

(c) $152 \times 2,040$

Solutions:

(a) $25 \times 111 = 25 \times 100 + 25 \times 10 + 25 \times 1 = 2,500 + 250 + 25$
$= 2,750 + 25 = 2,775$

(b) $22 \times 233 = 20 \times 233 + 2 \times 233 = 10 \times 2 \times 233 + 466 = 10 \times 466$
$+ 466 = 4660 + 466 = 5,126$

(c) $152 \times 2,040 = 152 \times 2,000 + 152 \times 40 = 152 \times 2 \times 1,000 + 152$
$\times 4 \times 10 = 304 \times 1,000 + 608 \times 10 = 304,000 + 6,080 = 310,080$

Multiplying by 5

When we were children playing hide-and-seek, the person who was "it" always seemed to count to 100 by fives: 5, 10, 15, 20, 25, 30, 35, 40, 45, 50, 55, 60, 65, 70, 75, 80, 85, 90, 95, 100. Other than helping you remember your fives multiplication table, there didn't seem to be any reason to do it that way rather than just counting to 20. We never counted by sevens, which was a table we could have used the practice on.

If your childhood experience didn't engrave the fives table in your memory, you may think multiplying by 5 mentally is challenging, but there is a shortcut: multiply by 10 (that is, just tack on a zero) and then divide by 2. To multiply 4×5, first multiply 4 by 10 to get 40, then divide by 2 to get 20. Ok, you probably knew that one anyway so let's try another.

Problem:

Multiply 28×5.

Solution:

$$28 \times 5 = 28 \times 10 \div 2 = 280 \div 2 = 140$$

You can, if you want to, divide by 2 first, and then multiply by 10, but the math is usually easier if you multiply by 10 first.

Problems:

Multiply each of the following:

(a) 86×5

(b) 124×50

(c) 377×5

Solutions:

(a) $86 \times 5 = 860 \div 2 = 430$

(b) $124 \times 50 = 124 \times 5 \times 10 = 124 \times 10 \div 2 \times 10 = 1,240 \div 2 \times 10$
    $= 620 \times 10 = 6,200$

(c) $377 \times 5 = 3,770 \div 2 = 1,885$

Two-Digit Number × 11

Multiplying a one–digit number by 11 is a piece of cake. Three times 11 is 33, 7 times 11 is 77, and 9 times 11 is 99. Multiplying a two-digit number by 11 is a little harder, but you already have one shortcut you can use.

Problem:

Multiply 42×11.

Solution:

$$42 \times 11 = 42 \times 10 + 42 \times 1 = 420 + 42 = 462$$

Here's the shortcut for the shortcut. To multiply a two-digit number, like 42, by 11:

1. Write the two digits with a space between them. 4_2

2. Add the digits together $(4 + 2 = 6)$ and put the result in the space. $42 \times 11 = 4\underline{6}2$

Problem:

Multiply 36×11

Solution:

$$36 \times 11 = 3\underline{9}6$$

Yes, there's a catch. Here it comes.

Problem:

Multiply 85 × 11

Solution:

Write the digits with a space between: 8_5

Add the digits 8 + 5 = 13. Oops! We've got a two-digit result, but only one space to fill in. So put the 3 in the space and "carry" the 1 over to the next place left. $\overset{1}{8}\underline{3}5$ = 935

Problem:

Multiply 79 × 11

Solution:

Write the digits with a space between: 7_9

Add the digits 7 + 9 = 16.

Put the 6 in the space and "carry" the 1 over to the next place left. $\overset{1}{7}\underline{6}9$ = 869

Problems:

Multiply each of the following:

(a) 186 × 5

(b) 32 × 11

(c) 67 × 11

(d) 49 × 50

(e) 29 × 110

Solutions:

(a) 930

(b) 352

(c) 737

(d) 2,450

(e) 3,190

Divisibility Tests

One of the ways to make division easier is to figure out before you do all the work if the numbers you're going to divide will work out nicely. We say a number is divisible by another if, when you do the division, there is no remainder. One number goes into the other "evenly." You get a whole number answer without any decimals or fractions.

For small numbers, knowing whether one number is divisible by another is just a matter of knowing your multiplication tables. You know 72 is divisible by 9 because you remember that 9 × 8 is 72, and you know 38 is not divisible by 6 because you recall that 6 × 6 is 36 and 6 × 7 is 42, so the 6-table jumps over 38.

For larger numbers, there are some tricks, as shown in Table 5.1.

You may be noticing that there is no trick for divisibility by 7 or 8, and no tricks for numbers larger than 10, and you may think that's a problem for large numbers. It's not as big a problem as you might think. First, if a number is not divisible by 2, it's not divisible by any even number, so that cuts the possibilities in half. If it's not divisible by 3, it won't be divisible by 6 or 9 or any multiple of 3. The same kind of thinking applies to 5: if your number's not divisible by 5, then it's not divisible by any multiple of 5. Finding numbers that do work,

Table 5.1   Divisibility Rules

A number is divisible by	If the number	Example
2	Ends in 0, 2, 4, 6, or 8	5,796 is divisible by 2 but 6,975 is not.
3	Has digits that add to a number divisible by 3	5,796 has digits that add to 5 + 7 + 9 + 6 = 27 and 27 has digits that add to 9. Nine is divisible by 3 so 5,796 is divisible by 3.
4	Can be divided by 2 and the result can be divided by 2 again	2,828 ÷ 2 = 1,414 and 1,414 ÷ 2 = 707 so 2,828 is divisible by 4.
5	Ends in 0 or 5	84,795 is divisible by 5 because it ends in 5. 71,110 is divisible by 5 because it ends in 0.
6	Is divisible by both 2 and 3	5,796 is divisible by 2 and 3 so it is divisible by 6. But 6,975 is divisible by 3 but not by 2 so it's not divisible by 6. And 2,828 is divisible by 2 but not by 3, so it's not divisible by 6.
9	Has digits that add to a number divisible by 9	27 has digits that add to 9 so it's divisible by 9 but 3,003 has digits that add to 6. It's divisible by 3 but not by 9.
10	Ends in 0	489,762,510 is divisible by 10 because it ends in 0.

and eliminating families of numbers that won't work, will give you the answer pretty quickly.

Suppose you're looking at the number 6,120 and you want to know if it's divisible by 11 or 17. There's no test for divisibility by 11 or by17, but what can you check for? It ends in 0 so it's divisible by 10 (or by 2 and 5, if you prefer). That means $6,120 = 612 \times 10$, and 612 is still divisible by 2 at least one more time.

$$6,120 = 612 \times 10$$
$$= 2 \times 306 \times 10$$
$$= 2 \times 2 \times 153 \times 10$$

Now, 153 is not divisible by 2 and therefore not divisible by any even number, so let's look at some odd numbers. Add the digits $1 + 5 + 3 = 9$, which is divisible by 3 (twice) or by 9. You may feel it's easier to divide by 3 and then by 3 again than to divide by 9, but it's your choice.

$$6,120 = 612 \times 10$$
$$= 2 \times 306 \times 10$$
$$= 2 \times 2 \times 153 \times 10$$
$$= 2 \times 2 \times 3 \times 51 \times 10$$
$$= 2 \times 2 \times 3 \times 3 \times 17 \times 10$$

You wanted to know if 6,120 was divisible by 11 or 17, and there's your answer. You can see the 17, and there's no sign of 11. Dividing 6,120 by 17 will give you 360 ($2 \times 2 \times 3 \times 3 \times 10$) with no remainder. But dividing 6,120 by 11 will leave a remainder; specifically, $6,120 \div 11 = 556$ and a remainder of 4.

SELF-TEST 5.3

1. Multiply by breaking into simpler problems:

   (a) 62×300 (b) 134×201 (c) $41 \times 5,000$

   (d) $86 \times 1,010$ (e) $29 \times 1,212$

2. Complete each multiplication, using whatever tricks you find helpful:

   (a) 57×11 (b) $81 \times 11,000$ (c) 23×5

   (d) 74×50 (e) 69×55

3. Use divisibility tests to help you determine if the first number is divisible by the second.

   (a) $198 \div 11$ (b) $459 \div 19$ (c) $520 \div 13$

   (d) $1,860 \div 31$ (e) $84,600 \div 47$

ANSWERS TO SELF-TEST 5.1

1. (a) 0.9   (b) 7   (c) 860

   (d) 56   (e) 1,020

2. (a) 400   (b) 2   (c) 1,700

   (d) 13   (e) 0.8

3. (a) 5,970,000   (b) 598,160,000   (c) 38,120,000,000

   (d) 1,172,900   (e) 5,100,000,000

ANSWERS TO SELF-TEST 5.2

1. (a) 0.001   (b) 0.4   (c) 90

   (d) 0.071   (e) 0.03

2. (a) 0.8   (b) 0.0014   (c) 0.037

   (d) 9.16   (e) 0.0005

3. (a) 9.432   (b) 0.00181   (c) 0.00059471

   (d) 65.42038   (e) 0.000000178635

1. (a) 18,600   (b) 26,934   (c) 205,000

   (d) 86,860   (e) 35,148

2. (a) 627   (b) 891,000   (c) 115

   (d) 3,700   (e) 3,795

3. (a) $198 = 2 \times 99 = 2 \times 9 \times 11$. 198 is divisible by 11. $198 \div 11 = 18$.

   (b) $459 = 3 \times 153 = 3 \times 3 \times 51 = 3 \times 3 \times 3 \times 17$. 459 is not divisible by 19.

   (c) $520 = 2 \times 260 = 2 \times 2 \times 130 = 2 \times 2 \times 13 \times 10$. 520 is divisible by 13. $520 \div 13 = 40$.

   (d) $1,860 = 186 \times 10 = 2 \times 93 \times 10 = 2 \times 3 \times 31 \times 10$. 1,860 is divisible by 31. $1,860 \div 31 = 60$.

   (e) $84,600 = 846 \times 100 = 2 \times 423 \times 100 = 2 \times 3 \times 141 \times 100 = 2 \times 3 \times 3 \times 47 \times 100$. 84,600 is divisible by 47. $84,600 \div 47 = 1,800$.

6 Positive and Negative Numbers

We're going to use a lot of what you learned in Chapter 2, "Essential Arithmetic," and some of Chapter 5, "Mental Math," during this chapter. There won't be any new operations, but there will be some new rules, because we'll be introducing some new numbers. Sometimes you encounter numbers that need to carry information not just about how much or how many, but also about direction. Did your favorite football team gain or lose yards? Did your weight go up or down? Did you just put $100 in your bank account or did you take $100 out?

1 WHAT ARE NEGATIVE NUMBERS?

If you went on a trip and a thief stole your wallet, you would end up with no money. What could be worse? You get into a friendly poker game, lose all your money, *and* owe your "friends" a couple of thousand dollars. Now that's getting into negative numbers.

The most frequent place to encounter a negative number is business losses. If your firm has sales of $500,000 and costs of $550,000, it has lost $50,000. If we think of a loss as a negative profit, then your business made a profit of −$50,000.

It may help to count down from positive to negative numbers. Positive numbers are ordinary everyday numbers: 5, 2, 16, 37, 9, 104. We could put + signs in front of them: +5, +2, +16, +37, +9, +104, but usually we don't bother. The only time we use these signs is when there are some negative numbers close by.

Let's start a countdown. This problem provides the first three numbers and you supply the next three.

Problem 1:

4, 3, 2, _____, _____, _____

Solution:

What did you get? You got 1 for the next number? So far, so good.

What's the next number? It's 0. And the one after 0? It's −1.

Here's another set.

Problem 2:

2, 1, 0, −1, _____, _____, _____

Solution:

I'll bet you wrote −2, −3, −4. Good, you're really getting into negative numbers now.

---

2 ADDING POSITIVE AND NEGATIVE NUMBERS

Negative numbers can be added, subtracted, multiplied, and divided just as positive numbers can. First, we'll do addition.

Add −3 and −5. Did you get −8? That's right. If you lost $3 to your poker-playing friends and then lost another $5, you'd be $8 in the hole.

Add +4 and −6. Did you get − 2? I hope so. You had been ahead by 4 and then lost 6, so now you're behind by 2.

Let's try adding −3, +4, −1, −5, and +2. You can add as you go across the line, or you may find it easier if you add all the positives together, add all the negatives together, and then add the positive total and the negative total. The choice is yours.

And the envelope, please. The answer is −3. If we add the positive numbers we get +6; if we add the negatives we get −9. Adding +6 and −9, we get −3. If you worked across the line −3 + 4 is +1. Add −1 to that and you have 0. Then −5 + 2 is −3.

Rules for adding positive and negative numbers quickly:

If the two numbers have the same sign (both + or both −), add the numbers and keep the sign.

$$8 + 5 = 13 - 7 + -4 = -11$$

If the two numbers have different signs (one + and one −), forget about the signs for a moment, subtract (yes, subtract) the smaller from the larger, and give your answer the sign of the number that looked larger.

−3 + 9 = 6 (9 looks bigger than 3. 9 − 3 = 6. 9 was positive so the answer is positive.)

−15 + 10 = −5 (15 looks bigger than 10. 15 − 10 = 5. 15 was negative so the answer is negative.)

3   SUBTRACTING POSITIVE AND NEGATIVE NUMBERS

Ready for some subtraction? Then here goes. Subtract -6 from -2. Suppose on first down, your team gained some yards, but on second down, they were thrown for a loss that wiped out what they did on the first down and two yards more. They're two yards behind the original line of scrimmage (where they started, if you're not a football fan), or -2. But the play is reviewed and the replay cancels (subtracts) the six-yard loss (-6). Where is your team? Your answer should be $+4$, four yards forward of the original line of scrimmage. They were at -2 but taking away the loss of 6, subtracting the -6, is equivalent to moving them ahead 6 yards. $-2 - -6 = -2 + 6 = 4$.

Imagine that you did your income tax return and figured out that you owed the government \$2,000, because your tax owed was \$2,000 more than your withholding. Withholding - Tax Owed = $-\$2,000$. Then the Internal Revenue Service informed you that you had made a \$6,000 mistake when calculating your withholding, so instead of being smaller than tax owed, it's bigger. That adds \$6,000 to your withholding, and your tax owed doesn't change. The IRS sends you a \$4,000 refund. $-\$2,000 + \$6,000 = \$4,000$.

Here's another problem. Subtract -3 from -10. What did you get? The right answer is -7. Here's the orthodox way of doing this: $-10 - (-3) = -10 + 3 = -7$. In other words, subtracting a -3 is really the same thing as adding a $+3$.

Rule for quick subtraction:

To subtract a number, add the opposite.

$$15 - 8 = 15 + (-8) = 7 \qquad -9 - 7 = -9 + (-7) = -16$$
$$11 - (-5) = 11 + 5 = 16 \qquad -8 - (-4) = -8 + (4) = -4$$

And finally, one last problem. Subtract -9 from -5. You get $-5 - (-9) = -5 + 9 = +4$.

SELF-TEST 6.1

1. Counting down: 5, 4, 3, _____, _____, _____, _____, _____
2. Counting up: -5, -4, -3, _____, _____, _____, _____, _____
3. Add -4 and -3.

4. Add −3, −1, +2, −6, and −2.

5. Add −5, +6, −2, −4, and +7.

6. Subtract −2 from +9.

7. Subtract −6 from −4.

8. Subtract +5 from −2.

If subtracting negative numbers is still not your strong suit, then take a look at the following box and then go on to Frame 4.

Subtracting a Negative Is Always a Plus

It may help to once again think of a negative number as a debt. Subtracting that debt is the same as adding that debt to your wealth. For instance, if you're $7 in debt and your creditor tells you to forget about the debt, you can state what happened this way:

$$-\$7 - (-\$7) = -\$7 + \$7 = 0$$

So now you're out of debt. When we subtract a minus, it's the same as adding a plus.

Problem 1:

How much is −2 subtracted from −6?

Solution:
$$-6 - (-2) = -6 + 2 = -4$$

Problem 2:

How much is 5 minus −4?

Solution:
$$5 - (-4) = 5 + 4 = 9$$

---

4   MULTIPLYING POSITIVE AND NEGATIVE NUMBERS

Ready to move up to a classier operation? Surprisingly, multiplying and dividing negative numbers is relatively easy. What's that? You're not convinced?

Okay, there is a catch. When you multiply two numbers, if just one of them has a negative sign, the answer is negative. Multiply -3×4. You should have gotten -12. Multiply 6×-3. Did you get -18?

When you multiply two positive numbers, you get a positive product. You've been doing that all your life. And when you multiply two negative numbers, you also get a positive product: $-3 \times -2 = +6$; $-5 \times -3 = +15$, or just plain 15.

Do you think you've gotten the hang of it? There's only one way to find out.

- Find the product of -7×2. It's -14. A negative times a positive gives us a negative.
- Find the product of -3×-8. It's 24. Two negatives give us a positive.
- And finally, how much is 5×6? It's 30. Two positives give us a positive.

Rules for quick multiplication and division:

- Multiply or divide the number parts as you always have.
- If the two numbers have the same sign, your answer is positive.
- If the two numbers have different signs, your answer is negative.

SELF-TEST 6.2

Find the products:

1. -1×-2 2. $-8 \times +7$

3. 6×9 4. 4×-7

5. 9×-9 6. -5×-10

7. 8×-8

If you are still having any trouble doing multiplication, you should definitely review the multiplication table (see Table 2.2). And it would probably be a good idea to take each of the self-tests in Chapters 3 through 5.

5 DIVISION WITH POSITIVE AND NEGATIVE NUMBERS

At last, we come to division with negative numbers—the moment we've all been waiting for. But first, we'll review our rules.

Division involving two positive numbers yields a positive quotient. Two negatives give us a positive quotient. And finally, when one number is positive and one is negative, the quotient will be negative. It doesn't matter whether the negative number is the dividend or the divisor. If there's one negative number in the division problem, the quotient is negative. If there are two negatives (or no negatives), the quotient is positive.

Is all this clear? Just remember, if both numbers have the same sign, the quotient is *positive*. If they have different signs, it's *negative*.

Divide −3 into 6. The answer is −2.

Divide −5 into −25. The answer is +5.

Divide 16 by −4. The answer is −4.

Divide −42 by 6. The answer is −7.

How much is 56 ÷ −8? The answer is −7.

How much is −63 ÷ −7? The answer is +9.

SELF-TEST 6.3

1. Divide −8 into 64.

2. Divide 7 into −35.

3. Divide −81 by −9.

4. Divide −80 by 10.

5. How much is −36 ÷ 4?

6. How much is 72 ÷ −9?

ANSWERS TO SELF-TEST 6.1

1. 2, 1, 0, −1, −2   2. −2, −1, 0, +1, +2

3. −7   4. − 0

5. +2   6. 11

7. +2   8. −7

ANSWERS TO SELF-TEST 6.2

1. $+2$ 2. -56

3. $+54$ 4. -28

5. -81 6. $+50$

7. -64

ANSWERS TO SELF-TEST 6.3

1. -8 2. -5

3. $+9$ 4. -8

5. -9 6. -8

7 Fractions

Did you ever wonder about the ÷ division sign? It's actually a tiny picture of a fraction. A fraction is a statement of division. The top number, called the numerator, is divided by the bottom number, called the denominator, to tell you how the two compare in size or what the part to whole relationship is. If you write the fraction $\frac{3}{5}$, you're representing the number you get when you divide 3 by 5 (which is less than 1). The fraction $\frac{1}{4}$ tells you that the part of a thing you have is one of four equal pieces.

When you're having a party, you might need two cakes, but what if you told the baker to give you parts of a few different ones, for variety. You might end up with $\frac{1}{2}$ of a pineapple upside-down cake, $\frac{1}{2}$ of a Black Forest cake, $\frac{1}{2}$ of a chocolate layer cake, and $\frac{1}{2}$ of a cheesecake. Add up these half cakes and see what you get. You would get $\frac{1}{2} + \frac{1}{2} + \frac{1}{2} + \frac{1}{2} = 2$ cakes. You've just done your first problem adding fractions. For you, it was a piece of cake—or, actually, four pieces of cake.

In this chapter we'll make use of what you learned in Chapter 2, "Essential Arithmetic," Chapter 3, "Focus on Multiplication," Chapter 5, "Mental Math," and a little bit of Chapter 4, "Focus on Division." Review if you feel you need to.

1 CHANGING THE LOOK OF FRACTIONS

One of the things that makes fractions a little more complicated than whole numbers is that it is possible to write many fractions that all represent the same number: $\frac{2}{4}, \frac{5}{10}, \frac{53}{106}$, and many others all can name the number we call one-half. That can sometimes be confusing, but it can also be useful. Let's look at the way to change the appearance of a number without changing its value. It's about the magic of the number 1.

Making a Whole Number Look Like a Fraction

Any whole number may be written over 1, for example, $\frac{47}{1}$. This does not change its value—any number divided by 1 *is* that number. This means that the arithmetic rules for fractions work for whole numbers too. Just write the whole number over 1 and it becomes a fraction.

Simplifying a Fraction

No one likes to work with large numbers when they can work with small ones. The fraction $\frac{630}{1,050}$ would intimidate most people. The trick is to look for and get rid of the "hidden ones." Can you think of a number that divides both 630 and 1,050? There are several, but most people think of 10 first. Write $\frac{630}{1,050}$ as $\frac{63 \times 10}{105 \times 10}$. Can you see that $\frac{10}{10} = 1$? That means $\frac{63 \times 10}{105 \times 10} = \frac{63}{105} \times 1 = \frac{63}{105}$, which is better. You might also realize that 63 and 105 can both be divided by 3 or by 7, or both. Get rid of some more 1s. $\frac{63}{105} = \frac{3 \times 3 \times 7}{5 \times 3 \times 7} = \frac{3}{5}$, which is much better.

Take the fraction $\frac{4}{10}$. That might leave *you* satisfied, but I can't tell you how much the sight of a 4 over a 10 would frustrate a mathematician, who would immediately simplify it to $\frac{2}{5}$. And it's not just fractions with even numbers on top and bottom that can be simplified—any fraction in which the numerator and denominator have a common divisor can be simplified.

Problem 1:

Express $\frac{20}{100}$ in its simplest form.

Solution:

$$\frac{20}{100} = \frac{2}{10} = \frac{1}{5}$$

In this case, we divided the top by 10 and the bottom by 10 (by cancelling the zeros), and then we divided the top and bottom by 2. Of course, if you immediately recognized that 20 went into 100 five times, you could have simply divided top and bottom by 20.

Problem 2:

Express $\frac{25}{500}$ in its simplest form.

Solution:

$$\frac{25}{500} = \frac{25 \times 1}{25 \times 20} = \frac{1}{20}$$

That's all there is to it.

Getting the Denominator You Want

On the other hand, sometimes you want to write your fraction using larger numbers, usually because there's a particular denominator (bottom number, also called the divisor) that you'd like it to have. This is still about "1." Multiplying a number by 1 doesn't change the value, and usually doesn't change the look, but what if you change the look of the 1? $\frac{3}{4} \times 1$ is still worth $\frac{3}{4}$ but suppose you want a denominator of 12. Write the one as $\frac{3}{3}$ and $\frac{3}{4} \times \frac{3}{3} = \frac{3 \times 3}{4 \times 3} = \frac{9}{12}$

    Shortcut: To change the look of a fraction without changing its value, multiply the numerator and denominator by the same number, or divide the numerator and denominator by the same number.

Problem 3:

Change $\frac{5}{8}$ to a denominator of 40.

Solution:
$$\frac{5}{8} = \frac{5 \times 5}{8 \times 5} = \frac{25}{40}$$

Problem 4:

Change $\frac{2}{9}$ to a denominator of 36.

Solution:
$$\frac{2}{9} = \frac{2 \times 4}{9 \times 4} = \frac{8}{36}$$

Changing a Mixed Number to a Fraction

A mixed number is a combination of a whole number and a fraction. If we write $2\frac{1}{5}$ meters, we're saying two meters and one-fifth of another meter. Rather than building another set of rules for mixed numbers, we change the mixed number to a fraction by writing the whole number over 1, changing to the same denominator as the fraction, and adding on the fraction:

$$2\frac{1}{5} = \frac{2}{1} + \frac{1}{5} = \frac{2}{1} \times \frac{5}{5} + \frac{1}{5} = \frac{10}{5} + \frac{1}{5} = \frac{11}{5}$$

    Shortcut: To change a mixed number to a fraction:

$$\frac{\text{denominator} \times \text{whole number} + \text{numerator}}{\text{denominator}}$$

Problem 5:

Change $6\frac{2}{3}$ to a fraction.

Solution:
$$6\frac{2}{3} = \frac{6}{1} + \frac{2}{3} = \frac{6 \times 3}{1 \times 3} + \frac{2}{3} = \frac{18}{3} + \frac{2}{3} = \frac{20}{3}$$

Problem 6:

Change $7\frac{3}{4}$ to a fraction.

Solution:

$$7\frac{3}{4} = \frac{7}{1} + \frac{3}{4} = \frac{7 \times 4}{1 \times 4} + \frac{3}{4} = \frac{28}{4} + \frac{3}{4} = \frac{31}{4}$$

2   ADDING FRACTIONS

Fractions, like whole numbers, can be added, subtracted, multiplied, and divided. First, we'll add them.

Problem 1:

Add $\frac{1}{2}$ and $\frac{1}{2}$.

Solution:

$$\frac{1}{2} + \frac{1}{2} = \frac{2}{2} = 1$$

That was an easy one. Here's something a bit harder.

Problem 2:

Add $\frac{1}{6}$ and $\frac{2}{6}$.

Solution:

$$\frac{1}{6} + \frac{2}{6} = \frac{3}{6} = \frac{1}{2}$$

Let's run the videotape for our instant replays. Everybody probably knows that $\frac{1}{2} + \frac{1}{2} = 1$. But what did we really do in terms of numerators (top numbers) and denominators (bottom numbers)? We added the numerators and wrote their sum over the denominator they shared.

$$\frac{1+1}{2} = \frac{2}{2} = 1$$

We did the same thing when we added $\frac{1}{6} + \frac{2}{6}$. We added the numerators over the common denominator:

$$\frac{1+2}{6} = \frac{3}{6} = \frac{1}{2}$$

Notice that we've rewritten this fraction with smaller numbers. Why? Mainly so that it is in its most recognizable form, or simplest form. For example, maybe *you* know that $\frac{176}{352} = \frac{1}{2}$, but not everyone else does. If you simplify your fractions to their simplest form, then anyone looking at your answers would know exactly how much they were.

When we add two or more fractions, we need to find the lowest common denominator. If you have 3 apples and 4 oranges, you don't have 7 apples, and you don't have 7 oranges. You need a new word to describe the 7 things you have, maybe 7 fruits. If you were to add $\frac{1}{5}$ and $\frac{1}{3}$, you would have to find a denominator that you can give to both fractions. It would be easy to change $\frac{1}{3}$ to a denominator of 6, but changing $\frac{1}{5}$ to a denominator of 6 would be difficult. A little thinking about multiplication facts will show you that the lowest common denominator is 15.

Before we can add $\frac{1}{5}$ and $\frac{1}{3}$, we need to express each with 15 as its denominator:

$$\frac{1}{5} = \frac{1 \times 3}{5 \times 3} = \frac{3}{15}$$

$$\frac{1}{3} = \frac{1 \times 5}{3 \times 5} = \frac{5}{15}$$

From here on out it's easy. Just add:

$$\frac{3}{15} + \frac{5}{15} = \frac{8}{15}$$

Problem 3:

Add $\frac{2}{3} + \frac{1}{6}$.

Solution:
$$\frac{2}{3} + \frac{1}{6} = \frac{2 \times 2}{3 \times 2} + \frac{1}{6} = \frac{4}{6} + \frac{1}{6} = \frac{5}{6}$$

Do you follow what we did here? We wanted to give both fractions a common denominator so we could add them. To convert $\frac{2}{3}$ into $\frac{4}{6}$, we multiplied the top and the bottom of the fraction by 2. Remember the law of arithmetic that allows this? We multiplied the top by 2 and the bottom by 2.

Here's another one.

Problem 4:

Add $\frac{1}{3} + \frac{2}{5}$.

Solution:

$$\frac{1}{3} + \frac{2}{5} = \frac{1 \times 5}{3 \times 5} + \frac{2 \times 3}{5 \times 3} = \frac{5}{15} + \frac{6}{15} = \frac{11}{15}$$

To repeat, we want the *lowest* common denominator to minimize the amount of simplifying later. And we want a common denominator so we'll be adding units of the same thing. Just as you can't add apples and oranges, you can't add thirds and quarters without finding their (lowest) common denominator, which happens to be 12.

And now for one more problem.

Problem 5:

Add $\frac{3}{4}$, $\frac{1}{3}$, and $\frac{1}{6}$.

Solution:

First find the lowest common denominator. In this case, it's 12 because 4 goes into 12, 3 goes into 12, and 6 goes into 12.

$$\frac{3}{4} + \frac{1}{3} + \frac{1}{6} = \frac{3 \times 3}{4 \times 3} + \frac{1 \times 4}{3 \times 4} + \frac{1 \times 2}{6 \times 2}$$

$$= \frac{9}{12} + \frac{4}{12} + \frac{2}{12} = \frac{15}{12} = 1\frac{3}{12} = 1\frac{1}{4}$$

As you've noticed, $\frac{15}{12}$ is greater than 1. When we convert it to $1\frac{3}{12}$, we are converting a fraction into a mixed number (a whole number and a fraction). The fraction $\frac{15}{12}$ tells us to divide 12 into 15, which gives us a quotient of 1 and a remainder of 3. We write that as $1\frac{3}{12}$. We'll encounter more mixed numbers in subsequent frames of this chapter and later chapters.

Are you getting the hang of it? Only you know for sure. If you've gotten the last two problems right, go on to Self-Test 1. Otherwise, please return to Frame 1.

SELF-TEST 7.1

1. $\frac{1}{4} + \frac{1}{5}$

2. $\frac{3}{10} + \frac{2}{5}$

3. $\frac{1}{2} + \frac{1}{3} + \frac{3}{4}$

4. $\frac{1}{6} + \frac{2}{3} + \frac{4}{9}$

3 SUBTRACTING FRACTIONS

There's absolutely nothing new when subtracting fractions except for a change of sign:

Problem 1:

$$\frac{5}{6} - \frac{1}{6}$$

Solution:

$$\frac{5}{6} - \frac{1}{6} = \frac{4}{6} = \frac{2}{3}$$

Problem 2:

$$\frac{1}{3} - \frac{1}{4}$$

Solution:

$$\frac{1}{3} - \frac{1}{4} = \frac{1 \times 4}{3 \times 4} - \frac{1 \times 3}{4 \times 3} = \frac{4}{12} - \frac{3}{12} = \frac{1}{12}$$

Problem 3:

$$\frac{3}{4} - \frac{2}{5}$$

Solution:

$$\frac{3}{4} - \frac{2}{5} = \frac{3 \times 5}{4 \times 5} - \frac{2 \times 4}{5 \times 4} = \frac{15}{20} - \frac{8}{20} = \frac{7}{20}$$

SELF-TEST 7.2

1. $\dfrac{3}{5} - \dfrac{1}{3}$ 2. $\dfrac{7}{8} - \dfrac{3}{4}$

3. $\dfrac{2}{3} - \dfrac{2}{5}$ 4. $\dfrac{1}{2} - \dfrac{1}{8}$

4 MULTIPLYING BY FRACTIONS

How do you find a fraction of a fraction, or a fraction of any number, for that matter? You may have heard that "of" means multiply, and that's generally true. This is a straightforward multiplication problem. So, let's set it up.

Problem 1:

How much is one-eighth of one-quarter?

Solution:

Write down one-eighth as a fraction. Then write down one-quarter, also known as one-fourth. Your fractions should look like this:

$$\frac{1}{8}, \frac{1}{4}$$

The final step is to multiply them:

$$\frac{1}{8} \times \frac{1}{4} = \frac{1 \times 1}{8 \times 4} = \frac{1}{32}$$

One nice thing about multiplying fractions is that it's not necessary to figure out a common denominator. But you'll find that when you multiply fractions, you can often simplify your result.

Here's another one.

Problem 2:

How much is one-third of one-seventh?

Solution:

$$\frac{1}{3} \times \frac{1}{7} = \frac{1 \times 1}{3 \times 7} = \frac{1}{21}$$

Now we'll get fancy.

Problem 3:

How much is one-half of two and a half? Hint: Set up both as fractions. Think of two and a half in terms of halves, or use the shortcut for changing a mixed number to a fraction.

Solution:

$$\frac{1}{2} \times 2\frac{1}{2} = \frac{1}{2} \times \frac{5}{2} = \frac{5}{4} = 1\frac{1}{4}$$

Are you getting the hang of it? I hope so. Try this one.

Problem 4:

Find one-third of seven.

Solution:

$$\frac{1}{3} \times \frac{7}{1} = \frac{7}{3} = 2\frac{1}{3}$$

The trick here is to write the number 7 in fraction form. Any number divided by 1 *is* that number: 7 divided by 1 is 7. We are allowed to write 7 as $\frac{7}{1}$. We put the 7 in fractional form so we can multiply it by $\frac{1}{3}$.

Problem 5:

Find two-fifths of nine.

Solution:

$$\frac{2}{5} \times \frac{9}{1} = \frac{18}{5} = 3\frac{3}{5}$$

Ready for another Self-Test? All right, then, here it comes.

SELF-TEST 7.3

1. Find two-thirds of one-sixth.

2. Find one-half of three and a half.

3. Find three-fifths of six.

4. Find three-quarters of twelve.

5 DIVIDING BY FRACTIONS

Let's get right into it. How much is one-half divided by one-fourth? Don't panic! Try phrasing the question differently. How many quarters in a half? Think of a pie or a cake, or anything you could cut up, divided into fourths. How many fourths make up a half? Can you see the answer?

Often the trick to doing a problem like this quickly and easily is to convert it into a multiplication problem. Remember that when you divide by a whole number, it's like multiplying by a fraction. Dividing by 2 is finding half of something. Dividing by 5 is multiplying by $\frac{1}{5}$. Dividing by a fraction is equivalent to multiplying by a number called the reciprocal of the fraction. To divide one-half by one-fourth, multiply one-half by the reciprocal of one-fourth. The reciprocal of a fraction is found by turning the fraction upside down. The reciprocal of $\frac{1}{4}$ becomes $\frac{4}{1}$.

Problem 1:

What is one-half divided by one-fourth?

Solution:

$$\frac{1}{2} \times \frac{4}{1} = \frac{4}{2} = 2$$

What gives us the right to convert a division problem into a multiplication problem just by converting one term into its reciprocal?

You could write it as a messy division problem of a fraction over a fraction and use our "hidden 1" trick.

$$\frac{^1/_2}{^1/_4} \times \frac{^4/_1}{^4/_1} = \frac{^1/_2 \times 4}{^1/_4 \times 4} = \frac{2}{1} = 2$$

That's actually sometimes the fastest way: just multiply the top and the bottom by the reciprocal of the bottom fraction. But how did we know to multiply by that number? Well, most people think of reciprocal as flipping the fraction upside down, and that is true, but the reciprocal of a number is actually defined as the number you multiply by to get an answer of one. $4 \times \frac{1}{4} = 1. \frac{3}{5} \times \frac{5}{3} = 1$ When you multiply the top and bottom by the reciprocal of the bottom fraction, you guarantee that you'll get a denominator of 1 when you multiply, and the number will be the answer to the division problem.

Problem 2:

How much is one-fifth divided by one-twentieth?

Solution:

$$\frac{1}{5} \times \frac{20}{1} = \frac{20}{5} = 4$$

Just remember to flip the fraction you're dividing by and you've set up your multiplication problem. You can remember it as KEEP-CHANGE-CHANGE: Keep the first number as it is, change the division sign to multiplication, and change the divisor to its reciprocal.

Now let's see how you do on Self-Test 7.4.

SELF-TEST 7.4

1. Find one-sixth divided by one-fourth.

2. Find two-fifths divided by two-thirds.

3. Find three-quarters divided by one-eighth.

4. Find one-third divided by two-fifths.

6   CANCELLING

Wouldn't you agree that it's a lot easier to multiply and divide relatively small numbers rather than large numbers? I thought so. Well, cancelling is a way of exchanging large numbers for small ones in multiplication problems. It's really just dividing, doing the simplifying ahead of time, and knowing the divisibility tests could come in handy.

In this problem, we'll do just that.

Multiply $\dfrac{15}{34} \times \dfrac{17}{5}$.

The basic rule for multiplication would tell you to do it this way: $\dfrac{15}{34} \times \dfrac{17}{5} = \dfrac{15 \times 17}{34 \times 5} = \dfrac{255}{170}$ but then you'll need to simplify.

$$\frac{15}{34} \times \frac{17}{5} = \frac{15 \times 17}{34 \times 5} = \frac{255}{170} = \frac{51 \times \cancel{5}}{34 \times \cancel{5}} = \frac{\cancel{17} \times 3}{\cancel{17} \times 2} = \frac{3}{2} = 1\frac{1}{2}$$

The easier way is to get the simplifying done before you multiply, like this:

$$\frac{15}{34} \times \frac{17}{5} = \frac{3 \times \cancel{5}}{2 \times \cancel{17}} \times \frac{\cancel{17}}{\cancel{5}} = \frac{3}{2}$$

We were able to divide 15 and 5 by 5, and to divide 34 and 17 by 17. Just remember it's always a numerator and a denominator. Cancel vertically or diagonally, never horizontally.

Please simplify and complete this problem: $\dfrac{5}{9} \times \dfrac{8}{25}$.

$$\frac{5}{9} \times \frac{8}{25} = \frac{\overset{1}{\cancel{5}}}{9} \times \frac{8}{\underset{5}{\cancel{25}}} = \frac{8}{45}$$

Exactly what did we do here? We recognized that 5 and 25 could both be divided by 5. To condense the work, we crossed out the 5 and wrote 1, and we crossed out the 25 and wrote 5. So $\frac{5}{9} \times \frac{8}{25}$ became $\frac{1}{9} \times \frac{8}{5}$.

What gives us the right to do this cancelling out? Remember the arithmetic rule that what you do to the top of a fraction, you must do to the bottom? You may claim that $\frac{5}{9}$ and $\frac{8}{25}$ are separate fractions, but I would be forced to disagree with you since they could be stated as:

$$\frac{5 \times 8}{9 \times 25}$$

Problem:

$$\frac{7}{18} \times \frac{9}{21}$$

Solution:

$$\frac{\overset{1}{7}}{\underset{2}{18}} \times \frac{\overset{1}{9}}{\underset{3}{21}} = \frac{1}{6}$$

One more Self-Test and we're out of here.

SELF-TEST 7.5

1. $\dfrac{45}{50} \times \dfrac{3}{9}$ 2. $\dfrac{36}{4} \times \dfrac{16}{6}$ 3. $\dfrac{12}{20} \times \dfrac{60}{24}$

ANSWERS TO SELF-TEST 7.1

1. $\dfrac{1}{4} + \dfrac{1}{5} = \dfrac{1 \times 5}{4 \times 5} + \dfrac{1 \times 4}{5 \times 4} = \dfrac{5}{20} + \dfrac{4}{20} = \dfrac{9}{20}$

2. $\dfrac{3}{10} + \dfrac{2}{5} = \dfrac{3}{10} + \dfrac{2 \times 2}{5 \times 2} = \dfrac{3}{10} + \dfrac{4}{10} = \dfrac{7}{10}$

3. $\dfrac{1}{2} + \dfrac{1}{3} + \dfrac{3}{4} = \dfrac{1 \times 6}{2 \times 6} + \dfrac{1 \times 4}{3 \times 4} + \dfrac{3 \times 3}{4 \times 3} = \dfrac{6}{12} + \dfrac{4}{12} + \dfrac{9}{12} = \dfrac{19}{12} = 1\dfrac{7}{12}$

4. $\dfrac{1}{6} + \dfrac{2}{3} + \dfrac{4}{9} = \dfrac{1 \times 3}{6 \times 3} + \dfrac{2 \times 6}{3 \times 6} + \dfrac{4 \times 2}{9 \times 2} = \dfrac{3}{18} + \dfrac{12}{18} + \dfrac{8}{18} = \dfrac{23}{18} = 1\dfrac{5}{18}$

ANSWERS TO SELF-TEST 7.2

1. $\dfrac{3}{5} - \dfrac{1}{3} = \dfrac{3 \times 3}{5 \times 3} - \dfrac{1 \times 5}{3 \times 5} = \dfrac{9}{15} - \dfrac{5}{15} = \dfrac{4}{15}$

2. $\dfrac{7}{8} - \dfrac{3}{4} = \dfrac{7}{8} - \dfrac{3 \times 2}{4 \times 2} = \dfrac{7}{8} - \dfrac{6}{8} = \dfrac{1}{8}$

3. $\dfrac{2}{3} - \dfrac{2}{5} = \dfrac{2 \times 5}{3 \times 5} - \dfrac{2 \times 3}{5 \times 3} = \dfrac{10}{15} - \dfrac{6}{15} = \dfrac{4}{15}$

4. $\dfrac{1}{2} - \dfrac{1}{8} = \dfrac{1 \times 4}{2 \times 4} - \dfrac{1}{8} = \dfrac{4}{8} - \dfrac{1}{8} = \dfrac{3}{8}$

ANSWERS TO SELF-TEST 7.3

1. $\dfrac{2}{3} \times \dfrac{1}{6} = \dfrac{2}{18} = \dfrac{1}{9}$ 2. $\dfrac{1}{2} \times \dfrac{7}{2} = \dfrac{7}{4} = 1\dfrac{3}{4}$

3. $\dfrac{3}{5} \times \dfrac{6}{1} = \dfrac{18}{5} = 3\dfrac{3}{5}$ 4. $\dfrac{3}{4} \times \dfrac{12}{1} = \dfrac{36}{4} = 9$

ANSWERS TO SELF-TEST 7.4

1. $\dfrac{1}{6} \times \dfrac{4}{1} = \dfrac{4}{6} = \dfrac{2}{3}$ 2. $\dfrac{2}{5} \times \dfrac{3}{2} = \dfrac{6}{10} = \dfrac{3}{5}$

3. $\dfrac{3}{4} \times \dfrac{8}{1} = \dfrac{24}{4} = 6$ 4. $\dfrac{1}{3} \times \dfrac{5}{2} = \dfrac{5}{6}$

ANSWERS TO SELF-TEST 7.5

1. $\dfrac{\overset{1}{\cancel{45}}^{5}}{\underset{10}{\cancel{50}}} \times \dfrac{3}{\underset{1}{\cancel{9}}} = \dfrac{3}{10}$

2. $\dfrac{\overset{6}{\cancel{36}}}{\underset{1}{\cancel{4}}} \times \dfrac{\overset{4}{\cancel{16}}}{\underset{1}{\cancel{6}}} = \dfrac{24}{1} = 24$

3. $\dfrac{\overset{1}{\cancel{12}}}{\underset{1}{\cancel{20}}} \times \dfrac{\overset{3}{\cancel{60}}}{\underset{2}{\cancel{24}}} = \dfrac{3}{2} = 1\dfrac{1}{2}$

<u>8</u> Decimals

Decimals are what we call our way of representing fractions in our base ten, or decimal system. Their "full name" is decimal fractions. We want to be able to write fractions in a decimal form so that they fit more comfortably into the system of arithmetic we're accustomed to. Much of what you need to know about arithmetic with decimals won't sound new, so let's get started.

1 ADDING AND SUBTRACTING WITH DECIMALS

There's not a lot new to say about adding and subtracting with decimals. Before you begin to add or subtract, you need to make sure that the decimal points of all the numbers are aligned one under another. Then add or subtract just as you did with whole numbers and place the decimal point in your answer right under all the others. If your problem includes whole numbers whose decimal point isn't showing, you can place the point and one or more zeros after the ones digit.

Problem 1:
Add 4.5 to 7.8.

Solution:

$$
\begin{array}{r}
\overset{1}{7}.8 \\
+\ 4.5 \\
\hline
12.3
\end{array}
$$

Problem 2:
Subtract 2.5 from 6.7.

Solution:

$$
\begin{array}{r}
6.7 \\
-\ 2.5 \\
\hline
4.2
\end{array}
$$

Problem 3:

Add 18.92 and 12.31.

Solution:

$$
\begin{array}{r}
^{1\,1} \\
18.92 \\
+\ 12.31 \\
\hline
31.23
\end{array}
$$

Problem 4:

Subtract 4.23 from 16.08.

Solution:

$$
\begin{array}{r}
^{5} \\
1\cancel{6}.^{1}08 \\
-\ 4.23 \\
\hline
11.85
\end{array}
$$

2 MULTIPLYING WITH DECIMALS

Multiplying with decimals is also a lot like whole number multiplication. The new feature is the decimal point, or more accurately, the location of the decimal point. We'll work out a simple problem and then I'll tell you the rule.

Problem 1:

$$1.5 \times 1.3$$

Solution:

$$
\begin{array}{r}
1.5 \\
\times 1.3 \\
\hline
45 \\
15\ \\
\hline
1.95
\end{array}
$$

This is really a two-part problem. The first part, the multiplication, should not present any difficulties. There's no need to line up decimal points, or carry the decimal points along as you work. Just multiply as though the decimal points weren't there. For a moment, pretend the decimal points weren't there and you were just multiplying 15×13. The multiplication part of this problem is exactly the same.

The second part of the problem is figuring out where to place the decimal point in the final answer. If you were multiplying 15×13, you'd probably realize that since $10 \times 10 = 100$, the answer will be bigger than 100, so 195 sounds reasonable. But for 1.5×1.3, you're multiplying two numbers that are bigger than 1 but smaller than 2. Your answer needs to be somewhere between $1 \times 1 = 1$ and $2 \times 2 = 4$.

Making an estimate of the result of a multiplication problem is a great help in deciding where the decimal point belongs. It may be all you need. You don't have to work at getting super close. More than 1 or less than 1? Between 1 and 10? Between 10 and 100? You just need to know what neighborhood you're in. That's harder with very large or very small numbers, of course, so there's also a rule to get you through the tougher ones.

Here's the rule: Count the number of digits to the *right* of the decimal point in each of the numbers being multiplied. Add those counts and place the decimal point in the answer so that there are that many digits to the right of the decimal point. For instance, 1.5 has one digit to the right of the decimal point, and 1.3 also has one digit to the right of the decimal point. That gives us two digits to the right of the decimal points. In our problem, the decimal point went after the 1 so that there were two digits, 9 and 5, after the decimal point.

Here's another one.

Problem 2:

$$\begin{array}{r} 2.53 \\ \times\, 8.6 \end{array}$$

Solution:

$$\begin{array}{r} 2.53 \\ \times\, 8.6 \\ \hline 1\ 518 \\ 20\ 24 \\ \hline 21.758 \end{array}$$

Estimate your answer. If we round both 2.53 and 8.6 down, we'd have $2 \times 8 = 16$. If we round up $3 \times 9 = 27$. Expect an answer between 16 and 27. We have two digits after the decimal in 2.53 and one digit after the decimal in 8.6. That's a total of three. We place the decimal in our answer so that there are three digits after it and get 21.758, which is between 16 and 27.

And now, another example.

Problem 3:

$$
\begin{array}{r}
41.59 \\
\times\, 76.84 \\
\end{array}
$$

Solution:

$$
\begin{array}{r}
41.59 \\
\times\, 76.84 \\
\hline
1\ 6636 \\
33\ 272 \\
249\ 54 \\
2911\ 3 \\
\hline
3195.7756 \\
\end{array}
$$

The numbers are a little larger so just estimate the neighborhood. $40 \times 80 = 3{,}200$. You should get a number in the thousands. The decimals don't change that. There are two digits after the decimal point in each number, for a total of four digits after the decimal point in the answer.

Are you getting these problems right? If not, redo them. If yes, then you're ready for Self-Test 8.1.

SELF-TEST 8.1

1.
$$
\begin{array}{r}
90.5 \\
\times\, 7.3 \\
\hline
\end{array}
$$

2.
$$
\begin{array}{r}
10.764 \\
\times\, 45.197 \\
\hline
\end{array}
$$

3.
$$
\begin{array}{r}
9.556 \\
\times\, 1.03 \\
\hline
\end{array}
$$

3   DIVIDING WITH DECIMALS

Like multiplying with decimals, dividing with decimals will look a lot like dividing with whole numbers, and once again the complicating factor is where to place the decimal point. If you're dividing a decimal by a whole number, it's simple. If your divisor is a decimal, or both numbers are decimals, you need a plan.

Decimals are, you'll recall, actually decimal fractions, so we'll borrow an idea from fractions to get the decimal point in the right place. You know that a fraction is a division problem, and that you can change its appearance by multiplying the numerator and denominator by the same number. If we were to write a division problem with a decimal divisor as a fraction, we could do this: $\frac{21}{0.5} = \frac{21 \times 10}{0.5 \times 10} = \frac{210}{5}$. That's a whole number division problem with the same answer as $21 \div 0.5$.

We'll start off with a very simple problem and work our way into the more complex.

Problem 1:

$$1.9\overline{)3.8}$$

Solution:

Think about multiplying $\frac{3.8}{1.9} \times \frac{10}{10}$

$$1.\underline{9}\,\overline{)3.\underline{8}} = 19\,\overset{2}{\overline{)38.0}}$$
$$\underline{-38}$$

Notice that all we did was move the decimal one place to the right for both numbers. That had the effect of multiplying both numbers by 10, changing the appearance of the "fraction" without changing its value.

Here's another one. Remember the fraction example. You have to multiply both decimals by the same number.

Problem 2:

$$4.52\overline{)15.1}$$

Solution:

$$4.\underline{52}\,\overline{)15.\underline{10}}$$

$$4.52\,\overset{3.34}{\overline{)1510.00}}$$
$$\overset{x\;x}{\underline{-1356}}$$
$$1540$$
$$\underline{-1356}$$
$$1840$$
$$\underline{-1808}$$
$$32$$

By now it should be clear that to divide we need to make the divisor into a whole number. We do that by moving the decimal point one or more places to the right. And what we do to the divisor, we need to do to the dividend, or the number under the division symbol.

SELF-TEST 8.2

1. $14.3\overline{)13}$

2. $.96\overline{)8.2}$

3. $.053\overline{)2}$

ANSWERS TO SELF-TEST 8.1

1.
```
    90.5
  × 7.3
  27 15
  633 5
  660.65
```

2.
```
     10.764
   × 45.197
     75348
     96876
   1 0764
  53 820
 430 56
 486.500508
```

3.
```
    9.556
  × 1.03
   28668
  9 5560
  9.84268
```

ANSWERS TO SELF-TEST 8.2

1. Round 0.909 to 0.91

$14.\underline{3}\overline{)13.\underline{0}}$

```
        0.909
  143 ) 130.000
         x x x
       −1287
        1300
       −1287
          13
```

2. Round 8.54 to 8.5.

$.\underline{96}\overline{)8.\underline{20}}$

```
         8.54
   96 ) 820.00
           x x
       −768
        520
       −480
        400
       −384
         16
```

3. Round 37.73 to 37.7.

$.\underline{053}\overline{)2.\underline{000}}$

```
          37.73
   53 ) 2000.00
             x x
       −159
        410
       −371
        390
       −371
        190
       −159
         31
```

9 Converting Fractions and Decimals

By now you probably realize that fractions and decimals are two sides of the same coin. Every fraction can be expressed as a decimal, and most decimals can be expressed as fractions. You probably already knew that $\frac{1}{4}$ of a dollar is $.25, that's where the 25-cent coin got the name quarter. How would you express $.50 as a fraction? It's half a dollar, or $\frac{1}{2}$ of a dollar. To do a lot of math problems, especially when you're doing algebra, you'll need to be able to convert fractions into decimals and decimals into fractions.

In this chapter, we'll be using what you learned in Chapter 7, "Fractions" and Chapter 8, "Decimals," but also Chapter 4, "Focus on Division," so you may want to look that over.

1 CONVERTING FRACTIONS INTO DECIMALS

Converting a fraction into a decimal is a simple problem of division. Divide the numerator by the denominator.

Problem 1:

Convert $\frac{1}{4}$ into a decimal.

Solution:

$$\frac{1}{4} = 4\overline{)1.00}^{.25}$$

A fraction is a division statement: numerator divided by denominator. Converting a fraction to a decimal simply means doing what it says.

Most often, fractions name a number that's less than one. When you divide a larger denominator into a smaller numerator, your quotient, or answer, will be less than 1, so 0 followed by a decimal point and decimal digits. The exception, which you saw with mixed numbers, is an improper fraction, where the numerator is larger than the denominator.

We'll try another.

Problem 2:

Convert $\frac{2}{5}$ into a decimal.

Solution:

$$\frac{2}{5} = 5\overline{)2.0}^{\;.4}$$

We'll do one last one.

Problem 3:

Convert $\frac{7}{10}$ into a decimal.

Solution:

$$\frac{7}{10} = 10\overline{)7.0}^{\;.7}$$

The problems in Self-Test 9.1 are like the ones you've just completed.

SELF-TEST 9.1

If you see a decimal repeating, round to the nearest thousandth (three decimal digits)

1. Convert $\frac{3}{4}$ into a decimal.

2. Convert $\frac{1}{6}$ into a decimal.

3. Convert $\frac{3}{8}$ into a decimal.

4. Convert $\frac{1}{12}$ into a decimal.

2   CONVERTING DECIMALS INTO FRACTIONS

Some decimals can easily be converted into fractions. For example, 0.3 becomes $\frac{3}{10}$. You can hear it in the decimal's proper mathematical name, three-tenths, (not "point three"). You can also get there by writing the decimal over one as a fraction and changing its appearance. Think of 0.3 as a fraction:

$$\frac{0.3}{1}$$

Any number may be written over 1 because that does not change its value—1 divided into any number *is* that number. Then we eliminate the decimal point by multiplying 0.3 by 10, and multiplying the 1 on the bottom by 10:

$$\frac{0.3 \times 10}{1 \times 10} = \frac{3}{10}$$

Want to try one on your own?

Problem 1:

Convert 0.57 into a fraction.

Solution:

$$0.57 = \frac{0.57}{1} = \frac{0.57 \times 100}{1 \times 100} = \frac{57}{100}$$

Problem 2:

Convert the following three-place decimal into a fraction: 0.833.

Solution:

$$0.833 = \frac{0.833}{1} = \frac{0.833 \times 1,000}{1 \times 1,000} = \frac{833}{1,000}$$

This method can be used for any decimal that terminates or ends. Now we can step back and generalize. When we're converting a decimal into a fraction, if that decimal has just one place (e.g., 0.3), then we remove the decimal point and divide by 10. If that decimal has two places (e.g., 0.57), then we remove the decimal point and divide by 100. And if that decimal has three places (e.g., 0.833), we remove the decimal point and divide by 1,000. In general, then, for every place we move the decimal point to the right (to remove it), we divide by a 1 followed by the same number of zeros as we moved the decimal point to the right.

SELF-TEST 9.2

1. Convert 0.13 into a fraction.

2. Convert 0.783 into a fraction.

3. Convert 0.9 into a fraction.

4. Convert 0.761 into a fraction.

3   SIMPLIFYING FRACTIONS

If you look back at all the problems we had in Frame 2 and in Self–Test 9.2, you were asked to convert decimals that ended with odd numbers into fractions: 0.3, 0.57, 0.833, 0.13, 0.783, 0.9, and 0.761. We deliberately avoided even numbers. Why? Because even numbers would likely leave us with fractions that would have to be simplified.

Now we'll combine two operations—converting decimals into fractions and then putting them in simplest form. Are you ready? Here comes the first one.

Problem 1:

Convert 0.45 into a fraction.

Solution:

$$0.45 = \frac{0.45}{1} = \frac{0.45 \times 100}{1 \times 100} = \frac{45}{100} = \frac{9}{20}$$

Here's one more.

Problem 2:

Convert 0.68 into a fraction.

Solution:

$$0.68 = \frac{.68}{1} = \frac{0.68 \times 100}{1 \times 100} = \frac{68}{100} = \frac{34}{50} = \frac{17}{25}$$

Now see how you do on Self-Test 9.3.

SELF-TEST 9.3

Convert each of these decimals into fractions in simplest form.

1. 0.75

2. 0.8

3. 0.65

4. 0.245

4 REPEATING DECIMALS

You might have noticed when you were converting fractions to decimals that sometimes, as you divided, you found the same remainder and the same digit in the quotient showing up again and again. This happened with $\frac{1}{6} = 0.1666\ldots$ and with $\frac{1}{12} = 0.08333\ldots$ It might have dawned on you that if you kept going, it would never change.

These repeating decimals often occur when the denominator is a multiple of 3, or 7, or another odd number, but for most everyday uses, you would do what we did in the answers above: round. Some common repeating decimals, like $\frac{1}{3} = 0.333\ldots$ and $\frac{2}{3} = 0.666\ldots$, show up so often that you get to know them by heart. Some repeating decimals start repeating right away, like $\frac{1}{3} = 0.333\ldots$ and some have a few other digits before the repeating starts, like $\frac{1}{12} = 0.08333\ldots$ Some repeat a single digit, like $\frac{2}{3} = 0.666\ldots$ while others, like $\frac{1}{7} = 0.142857142857\ldots$, have a group of digits that repeat.

You're unlikely to be asked to change a repeating decimal to a fraction. If it does come up, there's a shortcut, at least for decimals that start repeating right away. Write the pattern, whether it's one digit or many, as the numerator of the fraction. Count the number of digits in the repeating pattern, and write the same number of nines for the denominator, and simplify.

$$0.666\ldots = \frac{6}{9} = \frac{2}{3} \qquad 0.090909\ldots = \frac{09}{99} = \frac{1}{11}$$

$$0.142857142857\ldots = \frac{142{,}857}{999{,}999} = \frac{47{,}619}{333{,}333} = \frac{15{,}873}{111{,}111}$$

$$= \frac{1{,}443}{10{,}101} = \frac{481}{3367} = \frac{1}{7}$$

(Whew!)

Problem 1:

Find the decimal form of $\frac{4}{11}$.

Solution:

$$\begin{array}{r} 0.3\ 6\ 3\ 6 \\ 11\overline{)\,4.0^7 0^4 0^7 0} \end{array}$$

$$\frac{4}{11} = 0.363636\ldots = 0.\overline{36}$$

When a digit or a group of digits repeats forever, we can write it once with a bar over the top to show that it is repeating.

Problem 2:

Find the decimal form of $\frac{1}{6}$.

Solution:

$$6\overline{)1.0^40^40^40^40}$$
$$0.1\ 6\ 6\ 6\ 6$$

$$\frac{1}{6} = 0.166666\ldots = 0.1\overline{6}$$

Problem 3:

Express 0.44444 … as a fraction.

Solution:

There is one digit (4) in the repeating pattern. $0.44444\ldots = \frac{4}{9}$.

Problem 4:

Express 0.09090909 … as a fraction.

Solution:

There are two digits (09) in the repeating pattern.

$$0.09090909\ldots = \frac{09}{99} = \frac{9}{99} = \frac{1}{11}$$

5   NON-TERMINATING DECIMALS

Every fraction can be written as a decimal, but there are decimals that cannot be written as fractions. Numbers which cannot be written as fractions are called irrational numbers, since numbers that can be written as fractions are called rational numbers, because fractions are ratios. (More on ratios in Chapter 10, "Ratios and Proportions.")

Irrational numbers are decimals that don't terminate but don't repeat. The decimal 0.3 terminates and is equal to $\frac{3}{10}$. The decimal 0.333...repeats and is equal to $\frac{1}{3}$. Both of those are rational numbers. But if I write 0.09091092093 ... that's a decimal that doesn't terminate, but doesn't repeat because that one digit keeps changing.

Where in the world would you find such a number? It doesn't seem like there would be very many but in fact there are. However, the ones that come up are things like the square root of 2 (more on square roots later in Chapter 15, "Solving Simple Equations") and the number we call Pi, that comes up when we work with circles (Chapter 13, "Circumference and Area"). Don't get worried. For practical purposes, we'll round them off.

SELF-TEST 9.4

1. Convert $\frac{5}{12}$ to a decimal.

2. Convert $\frac{13}{30}$ to a decimal.

3. Convert 0.8888 ... to a fraction.

4. Convert 0.454545 ... to a fraction.

5. Explain how you know 0.1121231234 ... can't be written as a fraction.

ANSWERS TO SELF-TEST 9.1

1. $\frac{3}{4} = 4\overline{)3.00}$ .75

2. $\frac{1}{6} = 6\overline{)1.0000}$ .1666   = .167

3. $\frac{3}{8} = 8\overline{)3.000}$ .375

4. $\frac{1}{12} = 12\overline{)1.0000}$ .0833   = .083

ANSWERS TO SELF-TEST 9.2

1. $0.13 = \dfrac{0.13}{1} = \dfrac{0.13 \times 100}{1 \times 100} = \dfrac{13}{100}$

2. $0.783 = \dfrac{0.783}{1} = \dfrac{0.783 \times 1,000}{1 \times 1,000} = \dfrac{783}{1,000}$

3. $0.9 = \dfrac{0.9}{1} = \dfrac{0.9 \times 10}{1 \times 10} = \dfrac{9}{10}$

4. $0.761 = \dfrac{0.761}{1} = \dfrac{0.761 \times 1,000}{1 \times 1,000} = \dfrac{761}{1,000}$

ANSWERS TO SELF-TEST 9.3

1. $0.75 = \dfrac{0.75}{1} = \dfrac{0.75 \times 100}{1 \times 100} = \dfrac{75}{100} = \dfrac{3}{4}$

2. $0.8 = \dfrac{0.8}{1} = \dfrac{0.8 \times 10}{1 \times 10} = \dfrac{8}{10} = \dfrac{4}{5}$

3. $0.65 = \dfrac{0.65}{1} = \dfrac{0.65 \times 100}{1 \times 100} = \dfrac{65}{100} = \dfrac{13}{20}$

4. $0.245 = \dfrac{0.245}{1} = \dfrac{0.245 \times 1,000}{1 \times 1,000} = \dfrac{245}{1,000} = \dfrac{49}{200}$

ANSWERS TO SELF-TEST 9.4

1. $0.416666\ldots$

2. $0.43333\ldots$

3. $\dfrac{8}{9}$

4. $\dfrac{5}{11}$

5. Although there are digits that are repeated, the pattern of digits keeps changing. There's 1, then 12, then 123, and so on. Decimals that continue forever but don't repeat a pattern cannot be written as fractions.

10 Ratios and Proportions

When someone says you're blowing things way out of proportion, what does that mean? Way out of proportion to what? What is a proportion? As we study this chapter, we will come to understand the meaning and implications of ratios and proportions.

Throughout this chapter, we'll be using ideas from Chapter 7, "Fractions," so make sure you're comfortable with that.

1 RATIOS

Before we can discuss proportions, we must examine ratios. If someone says, I'll give you three to one on my team beating your team in the World Series, what does that person mean?

Three to one, or 3:1, is a ratio. If someone gives you 3 to 1 on a bet, it's his $3 to your $1. For non-gamblers, imagine you put your $1 and he puts his $3 into a pot. If you win, you get the pot of $4, so you have your dollar back and you win his $3. If he wins, he gets back his $3 and wins your $1.

A ratio is a way of comparing numbers, comparing the sizes of things, in a way that's based on division. Three to one, in the World Series example, suggests that your friend feels his chance of winning is three times as large as yours.

We use ratios all the time. A map may have a scale of 1 inch to 1 mile. If two points on the map are 3 inches apart, they're actually 3 miles apart.

There are 12 inches in a foot, or 12:1. There are two pints in a quart, or 2:1. And there are four quarts in a gallon, or 4:1. The two vertical dots, or colon, stand for the word "to." So, we can express a ratio as 3:1, or 3 to 1.

While the use of the word "to" or the *a:b* notation are common ways to represent ratios, you should also know that any ratio can also be written as a fraction. Twelve inches in one foot can be 12:1 or $\frac{12}{1}$. Ratios are comparisons based on division and fractions are statements of division. The additional piece you'll want to be on the lookout for is that ratios generally talk about quantities that include units: 12 *inches* in 1 *foot*, for example. Keeping them straight will be essential to solving problems.

You should also be careful to look for the information that puts the numbers in context. Are you comparing inches to miles for a map scale or inches to feet for a conversion? Are you comparing the number of days you ate ice cream to the number of days you didn't eat ice cream? Or the number of days you ate ice cream to the number of days in the year? Or the number of days you ate ice cream to the number of days you ate kale? It will matter.

Problem 1:

What is a team's ratio of wins to losses if they won 14 of 16 games?

Solution:

$$14:2 = 7:1$$

If you said 14:16, you read the numbers but didn't get the context. The question asked for the ratio of wins to losses, not to total games. You need to realize they lost $16 - 14 = 2$ games. Why not leave it at 14:2? There's nothing wrong with 14:2 mathematically, but by convention, ratios are expressed in their simplest form, just as fractions are expressed in simplest form. You get to simplest form of a ratio the same way you get to simplest form of a fraction, which is one reason some people prefer to write them in fraction form. The advantage of simplest form is that it aids us in comparing different ratios. It's a lot easier to make comparisons using 3:1 than 192:64.

Problem 2:

What is the ratio of ounces to pounds? (Hint: It helps to know that there are 16 ounces in every pound!)

Solution:

$$16:1$$

SELF-TEST 10.1

1. If someone offers you 8 to 1 odds on the championship boxing match, how much money would you have to put up to win $40?

2. What is the ratio of inches per yard?

3. What is the ratio of quarts to ounces?

4. A map has a scale of $1\frac{1}{2}$ inches to the mile and the distance between two points on the map is 9 inches. What is the distance in actual miles?

5. If an employee was out sick on 6 of 96 workdays, what is his ratio of sick days to days worked?

6. If a runner worked out 32 of the last 40 days, what is her ratio of days she worked out to days she did not work out?

2 PROPORTIONS

We're ready for proportions. What exactly is a proportion? It's a statement that two ratios are equal. By saying that two ratios are equal, we're saying that the relationship between the first two numbers is the same as the relationship between the second pair of numbers.

Imagine you had an old photo, maybe of a family member or a place that you visited once, and you'd like to enlarge it to frame it and hang it on your wall. The photo measures 3 inches by 5 inches. Its dimensions are in ratio 3:5. What do you think would happen if you made a copy of the photo that measured 6 inches by 20 inches? That copy would have dimension in ratio 6:20 or 3:10. Your picture would be distorted because the copy is not in proportion to the original.

Here's a proportion: 1 is to 2 as 3 is to 6, so 1:2 = 3:6. In this case, 1 has the same relation to 2 that 3 has to 6. And what exactly is that relation? What do you think? You're right: 1 is half the size of 2, and 3 is half the size of 6. Or, alternately, 2 is twice the size of 1, and 6 is twice the size of 3.

You could also say that a proportion is a statement that two fractions are equal, like saying $\frac{5}{8} = \frac{35}{56}$ or $\frac{24}{180} = \frac{2}{15}$. If you're thinking these statements look like what we did when we changed the look of a fraction without changing its value, you're absolutely right. We made proportional changes to the numerator and denominator.

We can use those connections back to things we know to help us find a missing number in a proportion. (When there's a missing number in any problem, it's common to pick a letter to take its place. We'll use x to stand for the missing number unless we tell you otherwise.)

Problem 1:

1:4 = 3:x. Find x.

Solution:

In the first ratio, the second number is 4 times the first, so in the second ratio, the missing number is $4 \times 3 = 12$. Write it as $\frac{1}{4} = \frac{3}{x}$ and think of it as changing the look of a fraction.

$$\frac{1 \times 3}{4 \times 3} = \frac{3}{x}. \text{ Either way, } x = 12.$$

Unfortunately, the numbers in a proportion aren't always so obvious. If we were to change that problem to 1:4 = x:27 (or $\frac{1}{4} = \frac{x}{27}$ if you prefer) you'd have to ask what you would multiply 4 by to get 27, and the answer isn't as pretty. You could do it, but here's another way, maybe an easier way.

A little vocabulary first. When a proportion is written $a{:}b = c{:}d$, the two numbers at the very beginning and the very end, a and d, are called the *extremes*. The two numbers in the middle, b and c, are the *means*. So, the proportion is

$$extreme{:}\ mean = mean{:}\ extreme$$

In any proportion:

The product of the means equals the product of the extremes.

In Problem 1 we have $1{:}4 = 3{:}x$. Multiplying the means, 4×3, gives us 12. Multiplying the extremes, $1 \times x$, gives us $1x$, or x. So we end up with $12 = x$. In the other example, $1{:}4 = x{:}27$, we multiply 4 times x and 1×27, and say 4 of those numbers we called x equal 27. If 4 of those unknown numbers make 27, you can find one of them by dividing 27 by 4.

$$x = \frac{27}{4} = 6\frac{3}{4}$$

When the proportion is written in fraction form, like $\frac{5}{12} = \frac{10}{x}$, the rule gets the name *cross-multiplying*. The extremes are the 5 and the x and the means are the 12 and the 10.

$$\text{Product of the means:} \quad \frac{5}{12} \nearrow \frac{10}{x}$$

$$\text{Product of the extremes:} \quad \frac{5}{12} \searrow \frac{10}{x}$$

$$5x = 120$$

$$x = \frac{120}{5} = 24$$

Let's do another one.

Problem 2:

Find x when $2{:}x = 5{:}15$.

Solution:

$$2{:}x = 5{:}15$$

$$5x = 2 \times 15$$

$$5x = 30$$

$$x = \frac{30}{5} = 6$$

How about this one?

Problem 3:

3 is to 5 as 9 is to what?

Solution:

The "what" we're looking for, we'll call x.

$$3{:}5 = 9{:}x$$
$$3x = 5 \times 9$$
$$3x = 45$$
$$x = \frac{45}{3} = 15$$

Now try this.

Problem 4:

4 is to what as 9 is to 30?

Solution:

$$4{:}x = 9{:}30$$
$$9x = 4 \times 30 = 120$$
$$9x = 120$$
$$x = \frac{120}{9} = \frac{40}{3} = 13\frac{1}{3}$$

The problems above told you pretty clearly what the proportion should be. In problems like these next few, that's not as obvious. Be careful to keep the same order in both ratios. If you start with apples : oranges, be sure your second ratio is apples : oranges and not oranges : apples.

Problem 5:

Budgeting advice often says that the ratio of your rent to your monthly income should be 1:4, how much should your monthly rent be if your income is $600 per month?

Solution:

Let rent = x

$$1{:}4 = x{:}\$600$$
$$4x = \$600$$
$$x = \frac{\$600}{4} = \$150$$

Problem 6:

If the ratio of Adam's age to his grandmother's age is 2:11 and Adam is 12, how old is his grandmother?

Solution:

Let grandmother's age $= x$

$$2{:}11 = 12{:}x$$

$$2x = 132$$

$$x = \frac{132}{2} = 66$$

Did you remember the shortcut for multiplying by 11 from Chapter 5, "Mental Math"? Are you ready for some apples and oranges?

Problem 7:

If 4 apples cost 39 cents, how much would 13 apples cost?

Solution:

Let $x =$ the cost of 13 apples

$$4{:}39 \text{ cents} = 13{:}x$$

$$4x = 507 \text{ cents}$$

$$x = \frac{507}{4} = 126.75 \text{ cents}$$

$$x = \$1.27$$

Writing in the unit "cents" helped here so that you didn't get to the end and think you were paying $126.75 for 13 apples. 127 cents is $1.27. You could also have used a proportion to find the cost of 1 apple, and then multiplied by 13. $4{:}39 = 1{:}x$ might be easier.

Problem 8:

If 7 oranges cost $1.05, how many oranges could be purchased for $3.00?

Solution:

Let $x =$ the number of oranges that could be purchased for $3

$$7{:}\$1.05 = x{:}\$3.00$$

$$\$1.05x = \$21$$

$$x = \frac{\$21}{\$1.05} = \frac{2100}{105} = \frac{420}{21} = 20$$

Problem 9:

A plane travels 2,000 miles in $4\frac{1}{2}$ hours. How long would it take that plane to travel 3,000 miles at that same rate of speed?

Solution:

Let x = length of time it would take the plane to travel 3,000 miles

$$4\frac{1}{2}:2,000 = x:3,000$$

$$2,000x = 4\frac{1}{2} \times 3,000 = 13,500$$

$$x = \frac{13,500}{2,000} = \frac{135}{20} = 6\frac{15}{20} = 6\frac{3}{4}$$

Problem 10:

In the city of San Francisco, approximately 1 in every 20 residents is Black or African American. Approximately how many Black or African American residents call San Francisco home if the city's population is 880,000?

Solution:

Let x = number of Black or African American people living in San Francisco

$$\frac{1}{20} = \frac{x}{880,000}$$

$$20x = 1 \times 880,000$$

$$x = \frac{880,000}{20} = 44,000$$

SELF-TEST 10.2

1. If a team wins 63 of 81 games, find their win-loss ratio.

2. 9 is to 12 as what number is to 40?

3. 6 is to 5 as 48 is to what number?

4. If you spend $1 out of every $10 of income on medical bills and if you paid $1,171 in medical bills, how much was your income?

5. $15:2 = x:10$. Find x.

6. $x:3 = 4:9$. Find x.

7. If 8 bananas cost $1.50, how much would 72 bananas cost?

8. If 2 pineapples cost $3.79, how many pineapples could be purchased for $22.74?

9. If you can drive 200 miles in $3\frac{3}{4}$ hours, how long would it take you to drive 500 miles at the same rate?

10. Approximately 580,000 of Chicago's residents claim Mexican heritage. If the population of Chicago is approximately 2,700,000 people, what is the ratio of people of Mexican heritage to those who do not claim that heritage?

11. If the ratio of people who identify themselves as a member of a religious group to those who do not is 4:1 in the city of Houston, how many Houstonians do not identify with a religious group if Houston had a population of 2 million?

ANSWERS TO SELF-TEST 10.1

1. $5

2. 36:1

3. 1:32

4. 6 miles

5. $6:90 = 1:15$

6. $32:8 = 4:1$

ANSWERS TO SELF-TEST 10.2

1. $63:18 = 7:2$

2. 30

3. 40

4. $11,710

5. 75

6. $x = 1\frac{1}{3}$

7. $13.50

8. 12

9. 9.375 hours

10. $580,000:2,120,000 = 29:106$

11. 400,000

<u>11</u> Solving Problems

Knowing how to perform the various operations you've learned in earlier chapters in this book is good. Knowing which one to perform in a real-life situation is better. That's why word problems keep coming back. And yes, the situations in word problems don't always sound like real life, but they do give you a chance to practice your skills in context. If you're impatient and want to get to other skills, skip around in this chapter and just make sure that you can set things up before you move on.

We'll be using a range of skills from Chapter 7, "Fractions," Chapter 8, "Decimals," and Chapter 10, "Ratios and Proportions." You may want to do the kind of conversions we discussed in Chapter 9, "Converting Fractions and Decimals," but you can probably avoid those if you try.

1 WORDS AND FRACTIONS

When you see some arithmetic written out with fractions in their number-divided-by-number form, what needs to be done is probably clear. When fractions are named, like one-fifth or three-sevenths, and the problem doesn't specifically spell out the operation to perform, it's up to you to translate.

Problem 1:
How much is one-quarter of one-third?

Solution:

$$\frac{1}{3} \times \frac{1}{4} = \frac{1}{12}$$

Problem 2:
Last year, Jason was 4 feet $8\frac{1}{8}$ inches tall. Now he is 5 feet $1\frac{1}{4}$ inches tall. How much did he grow over the last year? (Hint: Convert feet into inches.)

Solution:
5 feet = 60 inches, 4 feet = 48 inches.

$$5 \text{ feet } 1\frac{1}{4} \text{ inches} - 4 \text{ feet } 8\frac{1}{8} \text{inches} = 61\frac{1}{4} - 56\frac{1}{8}$$

$$= 61\frac{1 \times 2}{4 \times 2} - 56\frac{1}{8}$$

$$= 61\frac{2}{8} - 56\frac{1}{8} = 5\frac{1}{8} \text{ inches}$$

Problem 3:

Last week, Max lost $3\frac{1}{2}$ pounds, Karen lost $2\frac{7}{8}$ pounds, and Sharon lost $1\frac{3}{4}$ pounds. How much weight did the three of them lose all together?

Solution:

$$3\frac{1}{2} + 2\frac{7}{8} + 1\frac{3}{4} = \frac{7}{2} + \frac{23}{8} + \frac{7}{4}$$

$$= \frac{7 \times 4}{2 \times 4} + \frac{23}{8} + \frac{7 \times 2}{4 \times 2}$$

$$= \frac{28}{8} + \frac{23}{8} + \frac{14}{8}$$

$$= \frac{65}{8} = 8\frac{1}{8} \text{pounds}$$

An alternate method would be to add the whole numbers and fractions separately:

$$3 + 2 + 1 = 6$$

$$\frac{1}{2} + \frac{7}{8} + \frac{3}{4} = \frac{4}{8} + \frac{7}{8} + \frac{6}{8} = \frac{17}{8} = 2\frac{1}{8}$$

$$6 + 2\frac{1}{8} = 8\frac{1}{8}$$

Problem 4:

John bought two sets of weights totaling $315\frac{1}{2}$ pounds. If the first set weighed $132\frac{3}{4}$ pounds, how much did the second set weigh?

Solution:

$$315\frac{1}{2} - 132\frac{3}{4} = 315\frac{1 \times 2}{2 \times 2} - 132\frac{3}{4}$$

$$= 315\frac{2}{4} - 132\frac{3}{4}$$

How can we subtract $\frac{3}{4}$ from $\frac{2}{4}$? After all, $\frac{3}{4}$ is larger than $\frac{2}{4}$. The trick is to borrow 1 from 315 and rewrite it as an additional $\frac{4}{4}$. Adding $\frac{4}{4}$ to $\frac{2}{4}$ gives us $\frac{6}{4}$. Here it is, step by step:

$$315\frac{1}{2} \quad = \quad 315\frac{2}{4} \quad = \quad 314\frac{6}{4}$$

$$\frac{-132\frac{3}{4}}{} \qquad \frac{-132\frac{3}{4}}{} \qquad \frac{-132\frac{3}{4}}{182\frac{3}{4}}$$

Problem 5:

Elizabeth bought $12\frac{1}{3}$ yards of material to make dresses. She used $3\frac{1}{2}$ yards on the first dress and $3\frac{7}{8}$ on the second. How much material was left over?

Solution:

You might want to take this one in two steps. How much fabric has Elizabeth used? And then, how much fabric is left? You could do it all in one problem but it can get confusing.

Add what she used:

$$3\frac{1}{2} + 3\frac{7}{8} = 3\frac{1 \times 4}{2 \times 4} + 3\frac{7}{8}$$

$$= 3\frac{4}{8} + 3\frac{7}{8}$$

$$= 6 + \frac{11}{8}$$

$$= 6 + 1\frac{3}{8} = 7\frac{3}{8}$$

Then subtract the $7\frac{3}{8}$ yards she used from the $12\frac{1}{3}$ yards she started with. You'll need to do some regrouping again.

$$12\frac{1}{3} - 7\frac{3}{8} = 12\frac{1 \times 8}{3 \times 8} - 7\frac{3 \times 3}{8 \times 3}$$

$$= 12\frac{8}{24} - 7\frac{9}{24}$$

$$= 11 + 1 + \frac{8}{24} - 7\frac{9}{24}$$

$$= 11 + \frac{24}{24} + \frac{8}{24} - 7\frac{9}{24}$$

$$= 11\frac{32}{24} - 7\frac{9}{24} = 4\frac{23}{24}$$

Problem 6:

How many strips of wood $\frac{5}{8}$ of an inch wide can be sawed off a 5-foot piece of wood?

Solution:

5 feet is 60 inches. $60'' \div \frac{5''}{8} = \frac{\overset{12}{\cancel{60}}}{1} \times \frac{8}{\underset{1}{\cancel{5}}} = 96$

2 WORD PROBLEMS WITH MONEY

What's different about word problems with money? Not all that much really, but when it's about money, people pay more attention, perhaps because it feels important. The arithmetic is the same. You'll need to keep dollars and cents straight, round to the nearest penny, and remember that since our money system is base ten, we use decimals rather than fractions when we talk about money. For example, $\$12\frac{3}{4}$ describes the same amount of money as $12.75, but the decimal is preferred.

Problem 1:

How much would it cost to buy $4\frac{2}{3}$ yards of cloth at $11 per yard?

Solution:
$$4\frac{2}{3} \times \$11 = \frac{14}{3} \times \frac{\$11}{1} = \frac{\$154}{3} = \$51.33$$

Problem 2:

If silver wire were sold for 40 cents an inch, how much would it cost to buy 2 yards of wire?

Solution:

Remember to convert: 2 yards = 6 feet = 72 inches.

$$72 \times \$.40 = \$28.80$$

Problem 3:

Farmer Jones bought $7\frac{1}{2}$ bales of hay for $30. How much did one bale cost?

Solution:

We could divide $30 by $7\frac{1}{2}$ and get our cost per bale:

$$\frac{\$30}{7\frac{1}{2}} = \frac{\$30}{7.5} = 7.5\overline{)\$30} = 75\overline{)\$300} \atop \begin{array}{r} 4 \\ -300 \end{array}$$

But here's an easier way:

$$\$30 \div 7\frac{1}{2} = \frac{\$30}{1} \div \frac{15}{2} = \frac{\$30}{1} \times \frac{2}{15} = \frac{\$\overset{2}{30}}{1} \times \frac{2}{\underset{1}{15}} = \$4$$

If you'd like to review this type of operation, you'll find it in Chapter 7, Fractions. It should be emphasized that there's no one correct method to solve most mathematical problems. Like accountants, mathematicians are most concerned with the bottom line.

Let's see how you do on this self-test.

SELF-TEST 11.1

1. How much is two-sevenths of one-third?

2. Gillian and Amanda went on diets. Together they lost $20\frac{3}{4}$ pounds. If Amanda lost $12\frac{1}{8}$ pounds, how much did Gillian lose?

3. If you walked 23 miles at an average speed of $13\frac{1}{4}$ minutes per mile, how long did it take you to walk the entire distance?

4. If gold were selling at $450 an ounce, how much gold could you buy for $7,875?

ANSWERS TO SELF-TEST 11.1

1. $\frac{1}{3} \times \frac{2}{7} = \frac{2}{21}$

2. $20\frac{3}{4} - 12\frac{1}{8} = 20\frac{3 \times 2}{4 \times 2} - 12\frac{1}{8} = 20\frac{6}{8} - 12\frac{1}{8} = 8\frac{5}{8}$ pounds

3. $\frac{23}{1} \times \frac{53}{4} = \frac{1,219}{4} = 304\frac{3}{4}$ minutes, or 5 hours, 4 minutes, and 45 seconds

4. 17.5 ounces

$$
450\overline{)7875} = 45\overline{)787.5}
$$

$$
\begin{array}{r}
17.5 \\
45\overline{)787.5} \\
-45 \\
\hline
337 \\
-315 \\
\hline
225 \\
-225 \\
\hline
\end{array}
$$

12 Area and Perimeter

How big is a lot that is 2,000 square feet? If carpeting cost $13 a square yard, how much would it cost you to carpet a room that was 18 feet by 24 feet? When you have completed this chapter, you'll be able to answer all kinds of questions involving the areas of rectangles and triangles.

As we work through this chapter, you'll see material from Chapter 2, "Essential Arithmetic," as well as some from Chapter 7, "Fractions," and Chapter 8, "Decimals." Most of what we do here will be multiplication and division, but none of it should get too complicated. Review what you think you might need.

1 AREAS OF RECTANGLES

A rectangle is a two-dimensional box. Three rectangles are pictured. Rectangle A is a square. Notice that its sides are of equal length. Rectangle B's length is a little longer than its width. Rectangle C's length is much longer than its width.

The area of a rectangle is its length times its width.

Problem 1:
How much is the area of Rectangle A?

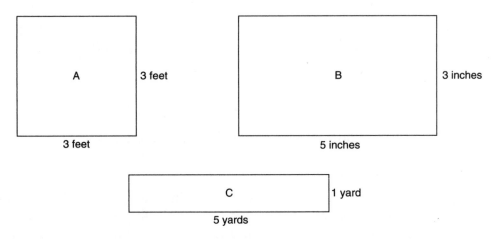

Solution:

$$\text{Area} = \text{length} \times \text{width}$$
$$= 3 \text{ feet} \times 3 \text{ feet}$$
$$= 9 \text{ square feet}$$

It is not enough to say that the area of Rectangle A is 9 or even 9 feet. Feet measure length. Square feet measure area. The area of Rectangle A is 9 square feet. If you multiply feet times feet, you will get square feet. Inches times inches will give you square inches and miles times miles will equal square miles. Feet times inches or inches time miles would make very little sense, so be sure that length and width are both in the same units.

Problem 2:

How much is the area of Rectangle B?

Solution:

$$\text{Area} = \text{length} \times \text{width}$$
$$= 5 \text{ inches} \times 3 \text{ inches}$$
$$= 15 \text{ square inches}$$

Problem 3:

And how much is the area of Rectangle C?

Solution:

$$\text{Area} = \text{length} \times \text{width}$$
$$= 5 \text{ yards} \times 1 \text{ yard}$$
$$= 5 \text{ square yards}$$

And now for something a little bit different.

Problem 4:

How many square inches are there in a square foot? (Imagine drawing a rectangle, a square actually, that measures 1 foot long and 1 foot wide, as in the following figure. It measures 12 inches by 12 inches.)

Solution:

We see from the following figure that the length and width of the square are both 12 inches. Therefore,

$$\text{Area} = \text{length} \times \text{width}$$

$$= 12 \text{ inches} \times 12 \text{ inches}$$

$$= 144 \text{ square inches}$$

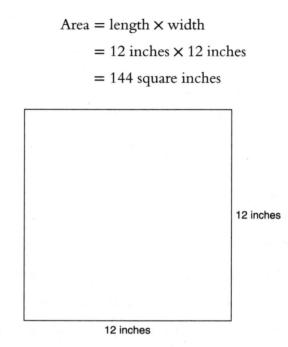

12 inches

12 inches

Algebraically we *could* say that the formula for the area of a square is its side squared, or s^2. (More on squares and related ideas coming up in Chapter 15, "Solving Simple Equations.") I'd rather treat the square as just another rectangle so that we can stick to one formula for the area of any rectangle: length × width.

Here's a similar question.

Problem 5:

How many square feet are there in a square yard?

Solution:

We see from the following figure that the length and width of the square are both 3 feet. Therefore,

$$\text{Area} = \text{length} \times \text{width}$$

$$= 3 \text{ feet} \times 3 \text{ feet}$$

$$= 9 \text{ square feet}$$

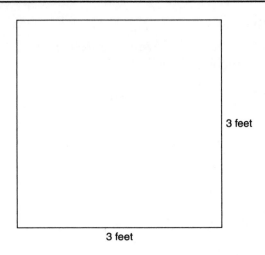

3 feet

3 feet

SELF-TEST 12.1

1. What is the area of a square that has sides of 6 inches?

2. What is the area of a rectangle that is 5 inches long and 4 inches wide?

3. What is the area of a rectangle that is 7 feet long and 3 feet wide?

2   WHEN WE USE AREA

Now, we'll do a few problems that you might be likely to encounter every day.

Problem 1:

A commercial building is renting for $10 per square foot. How much would it cost to rent an office that is 50 feet × 100 feet?

Solution:

$$\text{Area} = \text{length} \times \text{width}$$
$$= 100 \text{ feet} \times 50 \text{ feet}$$
$$= 5,000 \text{ square feet}$$

$$5,000 \text{ square feet} \times \$10 = \$50,000$$

Problem 2:

You buy a carpet that sells for $12 per square yard. If your room is 12 feet by 9 feet, how much will you spend?

Solution:

$$Area = length \times width$$
$$= 9 \text{ feet} \times 12 \text{ feet}$$
$$= 108 \text{ square feet}$$

There are 9 square feet in a square yard (3 feet × 3 feet = 9 square feet) and 108 square feet divided by 9 square feet per square yard tells you that your room is 12 square yards. So 12 square yards × $12 per square yard = $144. (In real life, you may pay more than that, if the carpet doesn't exactly fit your space and has to be cut and pieced together, leading to some waste. But this calculation gives you a good estimate.)

You could also do this problem by first changing the measurements of your room to yards: 9 feet = 3 yards, 12 feet = 4 yards. Then find the area as 3 × 4 = 12 square yards. If converting to yards is going to give you mixed numbers or messy decimals, you might not want to convert units until you must, but in real life, when you buy carpet, you'll be rounding measurements anyway.

Problem 3:

Manhattan's Central Park is approximately $2\frac{1}{2}$ miles long and $\frac{1}{2}$ mile wide. What is its area?

Solution:

$$Area = length \times width$$
$$= \frac{5}{2} \times \frac{1}{2}$$
$$= \frac{5}{4} = 1\frac{1}{4} \text{square miles}$$

Can you work backwards? If you know the area and the width, can you find the length? Of course you can! Just put the numbers you have into the right spots in the formula and divide.

Problem 4:

What is the length of a rectangular lot that has an area of 2,000 square feet and is 40 feet wide?

Solution:

$$\text{Area} = \text{length} \times \text{width}$$

$$2{,}000 \text{ square feet} = \text{length} \times 40 \text{ feet}$$

$$\frac{2{,}000 \text{ feet} \times \text{feet}}{40 \text{ feet}} = \text{length}$$

$$50 \text{ feet} = \text{length}$$

SELF-TEST 12.2

1. What is the area in square feet of a room that is 3 yards long and 2 yards wide?

2. If land were selling for $100 per square foot, how much would you have to pay for a square-shaped piece of land 80 feet long?

3. If a room is 30 feet by 30 feet, how much would it cost for wall-to-wall carpeting at $15 per square yard?

4. If a room were 3,000 square feet with a width of 15 feet, what is its length?

5. How many square inches are in a square yard?

PERIMETERS OF RECTANGLES

A perimeter is a border or outer boundary. You can find the perimeter of a rectangle (or, for that matter, a triangle, a pentagon, an octagon, etc.) by adding up the lengths of all the sides. For convenience, the perimeter of a rectangle can be found by using the formula:

Perimeter = (2 × length) + (2 × width) or Perimeter = 2 × (length + width)

Problem 1:

How much is the perimeter of the following figure?

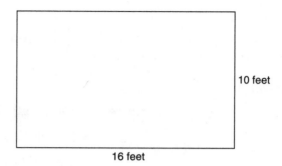

Solution:

$$\text{Perimeter} = (2 \times \text{length}) + (2 \times \text{width})$$
$$= (2 \times 16 \text{ feet}) + (2 \times 10 \text{ feet})$$
$$= 32 \text{ feet} + 20 \text{ feet}$$
$$= 52 \text{ feet}$$

Problem 2:

What is the perimeter of a rectangular field if its length is 100 yards and its width is 30 yards?

Solution:

$$\text{Perimeter} = 2 \times (\text{length} + \text{width})$$
$$= 2 \times (100 \text{ yards} + 30 \text{ yards})$$
$$= 2 \times 130 \text{ yards}$$
$$= 260 \text{ yards}$$

And now for something just a little bit different.

Problem 3:

What is the perimeter of a square whose side is 6 inches?

Solution:

$$\text{Perimeter} = (2 \times \text{length}) + (2 \times \text{width})$$
$$= (2 \times 6 \text{ inches}) + (2 \times 6 \text{ inches})$$
$$= 12 \text{ inches} + 12 \text{ inches}$$
$$= 24 \text{ inches}$$

Of course, we could have taken a shortcut by just multiplying 6 inches by 4, since the length and width of squares are equal.

Problem 4:

How much would it cost to build a fence around a lot that is 200 feet long and 80 feet wide if fencing cost $8 a foot?

Solution:

$$\text{Perimeter} = 2 \times (\text{length} + \text{width})$$
$$= 2 \times (200 + 80)$$
$$= 2 \times 280 \text{ feet}$$
$$= 560 \text{ feet}$$
$$\text{Cost} = 560 \times \$8$$
$$= \$4,480$$

Problem 5:

How much would it cost to build a wall around the perimeter of a lot that is 60 feet long and 30 feet wide, if it cost $20 to build each foot of the wall?

Solution:

$$\text{Perimeter} = (2 \times \text{length}) + (2 \times \text{width})$$
$$= (2 \times 60 \text{ feet}) + (2 \times 30 \text{ feet})$$
$$= 120 \text{ feet} + 60 \text{ feet}$$
$$= 180 \text{ feet}$$
$$\text{Cost} = 180 \text{ feet} \times \$20$$
$$= \$3,600$$

SELF-TEST 12.3

1. Find the perimeter of a field that is 500 feet long and 70 feet wide.

2. Find the perimeter of a square lot whose side is 40 feet.

3. How much would it cost to put a fence around a field that is 40 yards long and 20 yards wide if fencing cost $5 a foot (not $5 a yard)?

4. How much would it cost to put fencing around a square lot whose side is 50 feet if fencing cost $8 a foot?

4   AREAS OF TRIANGLES

Every triangle can be thought of as half of a four-sided figure. Draw a line from a corner to the opposite corner of a figure like a rectangle and you create two identical triangles. The area of the rectangle is length × width. In a triangle, we call the measurements the base and the height. (We could call them base and height in a rectangle too; it's just tradition.) The area of a rectangle is length × width (or base × height.) The area of a triangle is *one-half* the base times the height. The great truth of this statement should become evident when we look at the following figure.

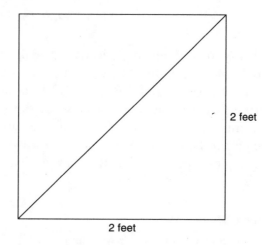

How much is the area of the square? Obviously, it's 4 square feet. Each triangle occupies half the area of the square. Now, how long is the base of the bottom triangle? It's 2 feet. And its height? Also 2 feet.

Problem 1:

Plug the numbers into the formula and see what you get for the area of the triangle:

$$\text{Area} = \frac{1}{2}\text{ base} \times \text{height}$$

Solution:

$$\text{Area} = \frac{1}{2}\text{ base} \times \text{height}$$

$$= \frac{1}{2} \times 2 \text{ feet} \times 2 \text{ feet}$$

$$= \frac{1}{2} \times 4 \text{ square feet}$$

$$= 2 \text{ square feet}$$

Problem 2:

If a triangle had a base of 4 feet and a height of 6 feet, how much is its area?

Solution:

$$\text{Area} = \frac{1}{2} \text{ base} \times \text{height}$$

$$= \frac{1}{2} \times 4 \text{ feet} \times 6 \text{ feet}$$

$$= \frac{1}{2} \times 24 \text{ square feet}$$

$$= 12 \text{ square feet}$$

When you cut a rectangle in half to make two triangles, it's easy to identify the base and the height. They are the two sides that meet at a corner of the rectangle, at a right angle. But not all four-sided figures are rectangles, so in some triangles, it's not so obvious what the base and height are. The plan is to pick one side of the triangle to call the base, and then measure the shortest distance from the opposite corner to your base, and call that the height. You'll see in the next problem that the base is always a side of the triangle, but the height is not.

Problem 3:

Using the following figure, find the area of Triangle *ACD*. In this picture, we're going to think of side $\overline{AD}$ as the base of the triangle, and we've drawn in $\overline{CB}$, which is not a side of $\triangle$ACD, but is the shortest distance from C to $\overline{AD}$. $\overline{CB}$ meets $\overline{AD}$ and makes right (90°) angles.

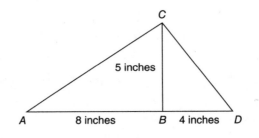

Let's try this one first by treating it as two triangles: one ($\triangle$ABC) with a base of 8 inches and a height of 5 inches, and one ($\triangle$BCD) with a base of 4 inches and a height of 5 inches.

Solution:

$$\text{Area of } \triangle ABC = \frac{1}{2} \text{ base} \times \text{height}$$

$$= \frac{1}{2} \times 8 \text{ inches} \times 5 \text{ inches}$$

$$= \frac{1}{2} \times 40 \text{ square inches}$$

$$= 20 \text{ square inches}$$

$$\text{Area of } \triangle BCD = \frac{1}{2} \text{ base} \times \text{height}$$

$$= \frac{1}{2} \times 4 \text{ inches} \times 5 \text{ inches}$$

$$= \frac{1}{2} \times 20 \text{ square inches}$$

$$= 10 \text{ square inches}$$

The area of $\triangle ACD$ = area of $\triangle ABC$ + area of $\triangle BCD$ = 20 + 10 = 30 square inches.

Divide and conquer is often a great strategy, but not always the fastest one. Here's the shorter method to find the area of $\triangle ACD$. The base of $\triangle ACD$ is 8 inches + 4 inches = 12 inches. It's height, from the top corner to the base, is 5 inches.

$$\text{Area of } \triangle ACD = \frac{1}{2} \text{ base} \times \text{height}$$

$$= \frac{1}{2} \times 12 \text{ inches} \times 5 \text{ inches}$$

$$= \frac{1}{2} \times 60 \text{ square inches}$$

$$= 30 \text{ square inches}$$

SELF-TEST 12.4

1. Using the figure, find the area of Triangle *EFG*.

2. Using the figure, find the area of Triangle *FGH*.

3. Using the figure, find the area of $\triangle EFH$.

4. If a triangle has a base of 15 inches and a height of 12 inches, how much is its area?

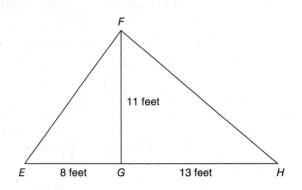

ANSWERS TO SELF-TEST 12.1

1. 6 inches × 6 inches = 36 square inches

2. 5 inches × 4 inches = 20 square inches

3. 7 feet × 3 feet = 21 square feet

ANSWERS TO SELF-TEST 12.2

1. 3 yards × 2 yards = 6 square yards
   6 × 9 = 54 square feet

2. 80 feet × 80 feet = 6,400 square feet
   6,400 × $100 = $640,000

3.   30 feet × 30 feet = 900 square feet
   100 square yards × $15 = $1,500

4.   Area = length × width
   3,000 square feet = length × 15 feet
         200 feet = length

5. 12 ×12 × 9 = 1,296 square inches

1. Perimeter = (2 × length) + (2 × width)
   = (2 × 500 feet) + (2 × 70 feet)
   = 1,140 feet
2. 4 × 40 feet = 160 feet

3. (2 × 40 yards) + (2 × 20 yards) = 120 yards
   $5 a foot = $15 a yard
   120 × $15 = $1,800
4. 4 × 50 feet = 200 feet
   200 feet × $8 = $1,600

1. Area $= \frac{1}{2} \times 8 \text{ ft} \times 11 \text{ ft} = \frac{1}{2} \times 88 \text{ sq ft} = 44 \text{ sq ft}$

2. Area $= \frac{1}{2} \times 13 \text{ ft} \times 11 \text{ ft} = \frac{1}{2} \times 143 \text{ sq ft} = 71.5 \text{ sq ft}$

3. Area $= \frac{1}{2} \times 21 \text{ ft} \times 11 \text{ ft} = \frac{1}{2} \times 231 \text{ sq ft} = 115.5 \text{ sq ft}$

4. Area $= \frac{1}{2} \times 15 \text{ in} \times 12 \text{ in} = \frac{1}{2} \times 180 \text{ sq in} = 90 \text{ sq in}$

13 Circumference and Area

What if a UFO landed one night on a beach, abducted a few midnight swimmers, and then took off? Suppose it left an imprint in the sand that measured 1,000 feet around. Just how big, then, was that UFO? If UFO had a circular base that measured 1,000 feet around, how wide was it?

In this chapter, we'll look at some of the same ideas as we did in the last chapter: measuring around the outside edge of a figure and measuring the area, or space inside the figure. This time, however, we'll be focused on circles, not rectangles or triangles. Circles don't have bases or heights or sides. They have no straight edges to measure, so we'll have to find some new ways to handle these questions. Along the way, we'll look at a very special number, the focus of mathematicians for centuries.

Like the previous chapter, this one will refer back to Chapter 2, "Essential Arithmetic," Chapter 7, "Fractions," and Chapter 8, "Decimals." The next time a UFO with a circular base lands in *your* neighborhood, you'll be ready to measure it.

1 CIRCUMFERENCE OF A CIRCLE

The *circumference* of a circle is the distance around that circle. It's related to the perimeter of a rectangle but we can't break it into lengths and widths. You could measure that distance by laying a bendable tape on the edge of a circle, or you could find the circumference indirectly by measuring across the widest point of the circle and doing a simple calculation. That measurement across the widest point of the circle is called the *diameter* of the circle. The diameter is actually a line segment, from one point on the circle to another directly opposite, which passes through the center of the circle, but we also call its measurement the diameter. The diameter is designated by the letter D, and is drawn in the circle shown in the following image.

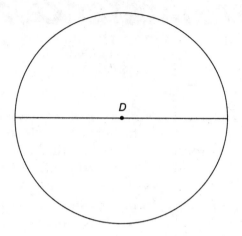

Many ancient mathematicians tried to find the relationship between the measurement of the diameter and the measurement of the circumference. Most of them estimated that the circumference was about three times the diameter. More than two millennia ago, the Greeks estimated that the circumference of a circle was about $3\frac{1}{7}$ times its diameter. A Babylonian tablet said it was about $3\frac{1}{8}$, an Egyptian source estimated about $3\frac{1}{6}$, and Chinese mathematicians calculated it as $\frac{355}{113}$ or $3\frac{16}{113}$. The approximate values most commonly used today are the $3\frac{1}{7}$ (or $\frac{22}{7}$) and 3.14. The Greeks designated the letter π (pronounced "pie") to stand for the number that you would get when you divide the circumference of a circle by its diameter. Modern mathematicians have determined that π is actually a decimal that never ends and have calculated it to thousands of decimal places. No one wants to have to remember all of them, so we use the approximate values or simply write the symbol π. The formula for the circumference of a circle is $C = \pi \times D$.

Problem 1:

If the diameter of a circle is 3 inches, what is the circumference of that circle?

Solution:

$$\text{Circumference} = \pi \times D$$

$$\approx \frac{22}{7} \times \frac{3}{1} \text{ inches}$$

$$\approx \frac{66}{7} \text{ inches}$$

$$\approx 9\frac{3}{7} \text{ inches}$$

We switched from an equal sign (=) to an approximately equal sign (≈) in the solution above when we replaced the symbol π with an approximate value. The exact value of the circumference could only be written as 3π. Any number we put in place of π is close, but not exactly, what π equals. Unfortunately, no one has yet made measuring tapes based on π, so we have to settle for an approximation. Accepting that reality, we won't always bother to change the equal sign.

Problem 2:

Find the circumference of a circle whose diameter is 14 feet.

Solution:

$$\text{Circumference} = \pi \times D$$

$$= \frac{22}{\underset{1}{7}} \times \frac{\overset{2}{\cancel{14}}}{1} \text{ feet}$$

$$= \frac{22}{1} \times \frac{2}{1} \text{ feet}$$

$$= 44 \text{ feet}$$

Now we'll add a new wrinkle by working backwards. If you know the diameter, you multiply by π to get the circumference. If you know the circumference, you DIVIDE by π to find the diameter.

Problem 3:

If the circumference of a circle is 8 inches, how much is the diameter?

Solution:

$$\text{Circumference} = \pi \times D$$

$$8 \text{ inches} = \frac{22}{7}D$$

$$D = 8 \div \frac{22}{7} = \frac{8}{1} \times \frac{7}{22} = \frac{56}{22} = 2\frac{6}{11} \text{ inches}$$

$$D \approx 2.5 \text{ inches}$$

Problem 4:

If the circumference of a circle is 12 feet, how much is the diameter?

Solution:

$$\text{Circumference} = \pi \times D$$

$$12 \text{ feet} = \frac{22}{7}D$$

$$12 \text{ feet} \div \frac{22}{7} = \frac{\overset{6}{\cancel{12}}}{1} \times \frac{7}{\underset{11}{\cancel{22}}} = \frac{42}{11} \text{ feet} = D$$

$$3.8 \text{ feet} \approx D$$

SELF-TEST 13.1

1. How much is the circumference of the circle in this figure?

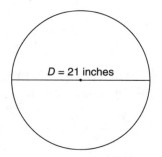

$D = 21$ inches

2. If the diameter of a circle is 9 feet, how much is its circumference?

3. If the circumference of a circle is 14 inches, how much is its diameter?

4. If the circumference of a circle is 9 feet, how much is its diameter?

2   THE AREA OF A CIRCLE

To find the area of a circle, we need to introduce one new term, the radius, which is represented by r. The radius is a line segment that connects the center of the circle to some point on the circumference. As with the diameter, we'll use the word radius to name the actual segment or its length. By definition, all radii (the plural of radius) of the same circle are equal.

Circle A in the following figure has a radius of 4 inches, as shown. If the radius of a circle is 4 inches, how much is the diameter of that circle? It must be 8 inches, since the diameter goes from a point on the circle, through the center, to another point on the circle. The diameter is composed of two radii, (radii is the Latin form of radiuses, and easier to pronounce) so the length of the diameter is twice the length of the radius. Circle B shows two radii of 3 inches each, which equals a diameter of 6 inches.

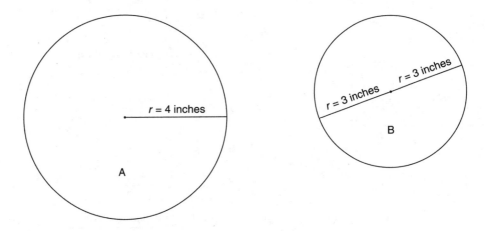

The formula for the area of a circle is $A = \pi r^2$. Writing r^2 is the same as saying $r \times r$. Let the square remind you that area is always measured in square units. Now there are those who would challenge the validity of the formula for the area of a circle: πr^2 (pronounced "pie r squared"). Once upon a time, mathematicians were held in the greatest of scorn and had trouble getting any form of employment. One such mathematician, who had a Ph.D. from a famous university and had written many learned textbooks, applied for a job as a dishwasher in a restaurant. The owner asked him for his qualifications and seemed unimpressed even after the mathematician had given him his résumé, along with an imposing list of publications.

"Tell me a mathematical formula," said the restaurant owner.

"πr^2," replied the mathematician.

"I am sorry," said the owner. "You have given me an incorrect formula. Pie are *round*; *cake* are square."

Here are some problems that involve π and squares.

Problem 1:

If the radius of a circle is 6 units, what is its area?

Solution:

$$\text{Area} = \pi r^2$$

$$= \frac{22}{7} \times \frac{6^2}{1}$$

$$= \frac{22}{7} \times \frac{36}{1}$$

$$= \frac{792}{7}$$

$$= 113\frac{1}{7} \text{ square units}$$

Problem 2:

What is the area of Circle B, shown earlier?

Solution:

$$\text{Area} = \pi r^2$$

$$= \frac{22}{7} \times \frac{(3 \text{ inches})^2}{1}$$

$$= \frac{22}{7} \times \frac{9 \text{ square inches}}{1}$$

$$= \frac{198 \text{ square inches}}{7}$$

$$= 28\frac{2}{7} \text{ square inches}$$

Problem 3:

If the diameter of a circle is 10 units, what is its area?

Solution:

Remember that the diameter is twice the radius, so the radius is half the diameter. If the diameter of the circle is 10 units, its radius measures 5 units.

$$\text{Area} = \pi r^2$$

$$= \frac{22}{7} \times \frac{5^2}{1}$$

$$= \frac{22}{7} \times \frac{25}{1}$$

$$= \frac{550}{7}$$

$$= 78\frac{4}{7} \text{ square units}$$

Problem 4:

If the diameter of a circle is 8 feet, what is its area?

Solution:

$$\text{Area} = \pi r^2$$

$$= \frac{22}{7} \times \frac{(4 \text{ feet})^2}{1}$$

$$= \frac{22}{7} \times \frac{16 \text{ square feet}}{1}$$

$$= \frac{352 \text{ square feet}}{7}$$

$$= 50\frac{2}{7} \text{ square feet}$$

SELF-TEST 13.2

1. If the radius of a circle is 4 inches, what is its area?

2. If the radius of a circle is 9 feet, what is its area?

3. If the diameter of a circle is 12 inches, what is its area?

4. If the diameter of a circle is 14 feet, what is its area?

ANSWERS TO SELF-TEST 13.1

1. $\text{Circumference} = \frac{22}{\underset{1}{\cancel{7}}} \times \frac{\overset{3}{\cancel{21}}}{1}$ inches

    $= 66$ inches

2. $\text{Circumference} = \frac{22}{7} \times \frac{9}{1}$ feet

    $= \frac{198}{7}$ feet

    $= 28\frac{2}{7}$ feet

3. $\text{Circumference} = \pi D$

    $14 \text{ inches} = \frac{22}{7}D$

    $\frac{14}{1} \div \frac{22}{7} = \frac{\overset{7}{\cancel{14}}}{1} \times \frac{7}{\underset{11}{\cancel{22}}} = \frac{49}{11}$ inches $= D$

    $4.5 \text{ inches} \approx D$

4. Circumference $= \pi D$

$$9 \text{ feet} = \frac{22}{7}D$$

$$\frac{9}{1} \div \frac{22}{7} = \frac{9}{1} \times \frac{7}{22} = \frac{63}{22} \text{ feet} = D$$

$$2.9 \text{ feet} \approx D$$

ANSWERS TO SELF-TEST 13.2

1. Area $= \pi r^2$

$$= \frac{22}{7} \times \frac{(4 \text{ inches})^2}{1} = \frac{22}{7} \times \frac{16 \text{ square inches}}{1} = \frac{352 \text{ square inches}}{7}$$

$$= 50\frac{2}{7} \text{ square inches}$$

2. Area $= \pi r^2$

$$= \frac{22}{7} \times \frac{(9 \text{ feet})^2}{1}$$

$$= \frac{22}{7} \times \frac{81 \text{ square feet}}{1} = \frac{1,782 \text{ square feet}}{7}$$

$$= 254\frac{4}{7} \text{ square feet}$$

3. Area $= \pi r^2$

$$= \frac{22}{7} \times \frac{(6 \text{ inches})^2}{1}$$

$$= \frac{22}{7} \times \frac{36 \text{ square inches}}{1}$$

$$= \frac{792 \text{ square inches}}{7}$$

$$= 113\frac{1}{7} \text{ square inches}$$

4. Area $= \pi r^2$

$$= \frac{22}{7} \times \frac{(7 \text{ feet})^2}{1}$$

$$= \frac{22}{7} \times \frac{49 \text{ square feet}}{1}$$

$$= \frac{22}{1} \times \frac{7 \text{ square feet}}{1}$$

$$= 154 \text{ square feet}$$

14   Percentages

Have you ever wondered what a percent is? You hear the word a lot, but what is a percent? It's a number that communicates the idea of a part of a whole, much the way a fraction or decimal does, but in a way that makes comparison easier. Which is larger, $\frac{6}{19}$ or $\frac{1}{3}$? Oh, wait, not sure? Which is larger, 32% or 33%? Simple, right? The term percent means "out of a hundred" and percents basically change fractions to have a denominator of 100, even if the numerator doesn't come out tidy.

A major corporation that produces lots of well-known cleaning products used to describe their soap as "99 and 44 one-hundredths percent pure." Very few people actually understood what that meant, but it sounded good. Nothing in life, not even soap, seems to be 100 percent pure (or 100% anything). Saying it's $99\frac{44}{100}$% pure (or 99.44%, if you prefer) sounds impressive but exactly *how* do you calculate percentages? That's the focus of this chapter.

We'll be relying heavily on ideas from Chapter 7, "Fractions," Chapter 8, "Decimals," Chapter 9, "Converting Fractions and Decimals," and short division from Chapter 4, "Focus on Division." Revisit those if you're feeling shaky about those skills, or if you run into difficulty during this chapter.

1   FRACTIONS, DECIMALS, AND PERCENTS

Percents can be expressed in fraction and decimal form—for instance, 1% is $\frac{1}{100}$ or one-hundredth. How much is 4%? Right—it's $\frac{4}{100}$. And how much is 43%? It's $\frac{43}{100}$. How much is $\frac{43}{100}$ in decimal form? The answer is 0.43. To summarize, $43\% = \frac{43}{100} = 0.43$. Usually, a percent is expressed by using a percent sign, so we can say that $76\% = \frac{76}{100} = 0.76$.

To convert a decimal to a percent, move the decimal point two places to the right and add a percent sign. For example, convert 0.23 to a percent. It's 23%. Convert 0.06 to a percent. It's 6%. Change 0.375 to a percent. The answer is 37.5%.

You probably noticed that all of the fractions so far had 100 as the bottom number. But we won't always be lucky enough to have 100 on the bottom.

Problem 1:

Convert $\frac{3}{10}$ into a percent.

Solution:

$$\frac{3 \times 10}{10 \times 10} = \frac{30}{100} = 30\%$$

Problem 2:

Change $\frac{1}{5}$ into a percent.

Solution:

$$\frac{1 \times 20}{5 \times 20} = \frac{20}{100} = 20\%$$

Though these are not so hard, there are some numbers that are more difficult to convert. It may not be possible or convenient to change them to a denominator of 100. Instead change the fraction to a decimal (remember Chapter 9?), move the decimal point two places to the right, and add a percent sign.

Problem 3:

Convert $\frac{7}{8}$ into a percent.

Solution:

$$8 \overline{)7.0^{6}0^{4}0} \quad \overset{.8\ 7\ 5}{} = 87.5\%$$

Ready for another?

Problem 4:

Convert $\frac{1}{6}$ into a percent.

Solution:

$$6 \overline{)1.0^{4}0^{4}0^{4}0} \quad \overset{.1\ 6\ 6\ 6}{} = 16.7\%$$

Rounding is common when we're changing to percents.

Problem 5:

Change $\frac{4}{3}$ into a percent.

Solution:

This is an improper fraction, a fraction greater than 1, so we'll get a percent greater than 100%.

$$\frac{1.\ 3\ 3\ 3\ 3}{3\overline{)4.^10^10^10^10}} = 133.3\%$$

For a practical example of these conversions, see the following box on how to calculate Social Security taxes.

How Your Social Security Tax Is Calculated

The federal government obtains funding for Social Security by collecting 6.2% of your paycheck from you and another 6.2% from your employer. If your salary is $300 a week, how much do *you* pay in Social Security tax?

Solution:

$$\begin{array}{r} \$300 \\ \times .062 \end{array}$$

or, to make it a little easier,

$$\begin{array}{r} .062 \\ \times \$\ 300 \\ \hline \$18.600 \end{array}$$

Every week, then, the government gets $18.60 in Social Security tax from your paycheck, and it collects another $18.60 from your employer. Both go into a fund from which you'll receive retirement benefits.

SELF-TEST 14.1

Convert these fractions into percents:

1. $\frac{15}{100}$

2. $\frac{36}{100}$

3. $\frac{122}{100}$

Convert these decimals into percents:

4. 0.75   5. 1.66   6. 0.05

Convert these fractions into percents:

7. $\frac{3}{20}$ 8. $\frac{4}{7}$ 9. $\frac{3}{8}$

---

2 PERCENT CHANGE

Somehow the idea of percent change—that is, percent increase or percent decrease—causes problems for folks, even if they have no difficulty changing fractions or decimals to percents. The key is in setting up the problem correctly, so make sure you pay close attention to that first step in these problems.

Problem 1:

If you were earning $400 a week and received a $10 raise, by what percent was your salary increased?

Solution:

$$\frac{\text{change}}{\text{original number}} = \frac{\$10}{\$400} = \frac{1}{40} = 2.5\%$$

The change, your raise, and the original number, what you were earning, were the only numbers you had for this problem, so this setup doesn't seem earthshaking. As we go along, remember its *change* over *original*, and make sure that's what you're using.

Problem 2:

If your IQ went from 85 to 140 after reading this book, by what percent did your IQ change?

Solution:

$$\frac{\text{change}}{\text{original number}} = \frac{55}{85} = \frac{11}{17} = 64.7\%$$

$$
\begin{array}{r}
.647 \\
17{\overline{\smash{\big)}\,11.000}} \\
\underline{-10\ 2} \\
80 \\
\underline{-68} \\
120 \\
\underline{-119} \\
1
\end{array}
$$

Notice that you weren't given the change. You had to calculate it by subtracting 140 – 85.

Problem 3:

A baseball team had a better year in 2019 than they had in 2018. In 2019 they won 70 games; in 2018 they won only 56. By what percentage did their wins increase?

Solution:

$$\frac{14}{56} = \frac{2}{8} = \frac{1}{4} = .25 = 25\%$$

If you got this right, go on to Self–Test 14.2. If you didn't, then you may be wondering where the 14 came from. If the team won 70 games in 2019 and 56 in 2018, then the change is 14. If you feel you need further review, go back to the beginning of Frame 2.

SELF-TEST 14.2

1. If you went on a diet and lost 35 pounds, by what percent did your weight decline if you now weigh 100 pounds?

2. If you grew from 5 feet 3 inches to 5 feet 7 inches, by what percent did your height increase?

3. If your pay was cut from $55,000 to $52,000, by what percent did your pay decline?

3 FAST PERCENT CHANGE

Pick a number, any number. Now triple it. By what percent did your number increase? Take your time. Calculate the percent. Don't just guess. What did you get? Three hundred percent? Sorry, that's incorrect. Did you remember to find the change, not the new amount?

As an example, let's use 100. Now let's triple it. We have 300. How much is the percent increase when you go from 100 to 300? It's the increase, in this case, 200, over the original, 100. That's a 200% increase. Whenever you go from 100 to a higher number, the percent increase is the difference between 100 and the new number: 300 − 100 = 200. Suppose you quadruple a number

$(400 - 100 = 300)$; the increase is also the difference between 100 and the new number. Pretty easy, huh?

We'll try one more. When you double a number, the percent increase is 100%. Use our original formula to calculate this percent change:

$$\frac{\text{change}}{\text{original number}}$$

If we double a number, in this case, doubling 100, we would get:

$$\frac{100}{100} = 100\%$$

What would the percent increase be if we went from 7 to 21? It would be 200%. How did we get this? We use the formula—change divided by the original number—or we remember that doubling is a 100% increase and tripling is a 200% increase.

Now I'm going to throw you a curveball. If a number—any number—were to decline by 100%, what number would you be left with? I'd really like you to think about this one.

What did you get? You should have gotten 0. That's right—no matter what number you started with, a 100% decline leaves you with 0.

Try the number 15. Now write down the formula for percentage change:

$$\frac{\text{change}}{\text{original number}}$$

Next, substitute into the formula:

$$-100\% = \frac{?}{15}$$

Since 100% is equal to 1, we have:

$$-1 = \frac{?}{15} \text{ or } ? = -15$$

In other words, if the change from the original number 15 is also 15, that means we went from the original number 15 all the way to 0.

SELF-TEST 14.3

1. A change from 10 to 50 represents a percent change of how much?

2. A change from 50 to 0 represents a percent change of how much?

3. A change from 20 to 50 represents a percent change of how much?

4. A 100% decline from 65 leaves us with how much?

4 PERCENT DISTRIBUTION

Are you eating too much red meat? You may be asking what that has to do with math. First, there are the pluses and the minuses. Red meat is a great source of protein; however, it's also a source of cholesterol and fat. It can be part of a balanced diet that gets protein from a variety of sources. Let's look at the distribution of protein sources in your diet by percent.

Problem 1:

If over the course of a week, you obtained 250 grams of protein from red meat, 150 from fish, 100 from poultry, and 50 from other sources, what percentage of your protein intake came from red meat and what percent came from each of the other sources?

red meat	250	grams
fish	150	grams
poultry	100	grams
other	50	grams
	550	grams

You now should be able to work this out for yourself. Hint: 550 grams = 100%.

Solution:

$$\text{red meat} = \frac{250}{550} = \frac{25}{55} = \frac{5}{11} = 45.5\%$$

$$\text{fish} = \frac{150}{550} = \frac{15}{55} = \frac{3}{11} = 27.3\%$$

$$\text{poultry} = \frac{100}{550} = \frac{10}{55} = \frac{2}{11} = 18.2\%$$

$$\text{other} = \frac{50}{550} = \frac{5}{55} = \frac{1}{11} = 9.1\%$$

$$\begin{array}{r} .4\ 5\ 4\ 5 \\ 11\overline{)5.0^6 0^5 0^6 0} \end{array}$$

$$\begin{array}{r} .2\ 7\ 2\ 7 \\ 11\overline{)3.0^8 0^3 0^8 0} \end{array}$$

$$\begin{array}{r} .1\ 8\ 1\ 8 \\ 11\overline{)2.0^9 0^2 0^9 0} \end{array}$$

$$\begin{array}{r} .09\ 09 \\ 11\overline{)1.00^1 00} \end{array}$$

Check:

$$
\begin{array}{r}
\overset{1}{} \\
45.5 \\
27.3 \\
18.2 \\
\underline{9.1} \\
100.1
\end{array}
$$

It's always a good idea to run a check of your work. The individual percentage shares should add up to about 100%. In this case, because of rounding, we ended up at slightly more than 100.0%.

Now we'll try another example.

Problem 2:

A family spent $7,000 on food, $4,000 on clothing, $5,500 on shelter, and $3,500 on other miscellaneous items. Find the percent of their expenses the family spent on each item.

food	$7,000
clothing	4,000
shelter	5,500
other	3,500
	$20,000

Solution:

$$\text{food} = \frac{7{,}000}{20{,}000} = \frac{7}{20} = .35 = 35\%$$

$$\text{clothing} = \frac{4{,}000}{20{,}000} = \frac{4}{20} = \frac{2}{10} = .2 = 20\%$$

$$\text{shelter} = \frac{5{,}500}{20{,}000} = \frac{55}{200} = \frac{11}{40} = 27.5\% \qquad 40\overline{)11.0} = \quad 4\overline{)1.1^30^20}\,^{.2\ 7\ 5}$$

(Note this maneuver we've just executed. We divided 40 by 10 and 11 by 10, making a long division problem into a short one.)

$$\text{other} = \frac{3{,}500}{20{,}000} = \frac{35}{200} = \frac{7}{40} \qquad 40\overline{)7.0} = \quad 4\overline{).7^30^20}\,^{.1\ 7\ 5} = 17.5\%$$

Check:
$$35.0\%$$
$$20.0$$
$$27.5$$
$$\underline{17.5}$$
$$100.0\%$$

SELF-TEST 14.4

Solve each problem. Use a calculator where appropriate.

1. A province has 1.9 million citizens who report their ethnicity as South Asian, 1.8 million as East Asian, 1.2 million as Black, and 2.7 million as Other. Find the percent share of each population group.

2. A college had 700 freshmen, 650 sophomores, 600 juniors, and 550 seniors. Find the percent share of each class.

3. A bank lent $200 million to consumers, $300 million to businesses, and $500 million to the government. Find the percent share of each type of loan.

5   TIPPING

One of the greatest social questions of all time is how much to leave as a tip in a restaurant. Etiquette experts may have suggested 15% but many people feel 20% is more appropriate. This percentage might vary, of course, with such factors as quality of service, friendliness of the waitstaff, ambiance, quality of the meal, and size of the check.

You probably don't want to pull out your calculator to determine the exact 15%, although many people do. So consider your alternatives.

In New York City, people often double the tax. The sales tax in New York City is actually 8.875%, so doubling the tax means they tip 17.75%, which is in the recommended range. Other cities will have different tax rates, so this tactic may not work, or you may be able to triple your tax.

Since multiplying the exact amount of the check by 0.15 or 0.20 in your head may be difficult, let's figure out a fast way of approximating the tip.

Start by rounding the amount of the bill to a round number. If your bill comes to $39.82, round it to $40. Next, take 10% of that number. Remember that 10% is $\frac{1}{10}$ of the total, so divide by 10 by moving the decimal point one place left or just cancel that last zero. Ten percent of $40 is $4. Do you want a 20% tip? Just double the 10% and leave your server $8. If you'd rather leave a

15% tip, remember that 5% is half of 10%, so 5% of $40 is half of $4, or $2. Add the 5% to the 10%: $2 + $4 = $6. A 15% tip is $6.

If your bill comes to $63.09, how do you calculate the tip? First, round the figure to $63; 10% comes to $6.30, and 5% is $3.15. If you want to leave a 15% tip, that's a total of $9.45. If you prefer a 20% tip, just double $6.30 and the tip is $12.60.

SELF-TEST 14.5

If you left a 15% tip, how much money would you leave on each of these restaurant checks?

1. $20.00

2. $29.50

If you left a 20% tip, how much money would you leave on each of these restaurant checks?

3. $45.00

4. $73.95

ANSWERS TO SELF-TEST 14.1

1. 15%

2. 36%

3. 122%

4. 75%

5. 166%

6. 5%

7. $\dfrac{3 \times 5}{20 \times 5} = \dfrac{15}{100} = 15\%$

8. $\begin{array}{r} .5\ 7\ 1\ 4 \\ 7\overline{)4.0^5 0^1 0^3 0} \end{array} = 57.1\%$

9. $\begin{array}{r} .3\ 7\ 5 \\ 8\overline{)3.0^6 0^4 0} \end{array} = 37.5\%$

ANSWERS TO SELF-TEST 14.2

1. $\dfrac{35}{135} = \dfrac{7}{27}$

$$\begin{array}{r} .259 \\ 27\overline{)7.000} \\ ^{x\,x} \\ \underline{-54} \\ 160 \\ \underline{-135} \\ 250 \\ \underline{-243} \\ 7 \end{array} = 25.9\%$$

2. 5 feet 3 inches = 63 inches;
    5 feet 7 inches = 67 inches

$\dfrac{4}{63}$

$$\begin{array}{r} .0634 \\ 63\overline{)4.0000} \\ ^{x\,x} \\ \underline{-378} \\ 220 \\ \underline{-189} \\ 310 \\ \underline{-252} \\ 58 \end{array} = 6.5\%$$

3. $\dfrac{\$3{,}000}{\$55{,}000} = \dfrac{3}{55}$

$$\begin{array}{r} .0545 \\ 55\overline{)3.0000} \\ \underline{-275} \\ 250 \\ \underline{-220} \\ 300 \\ \underline{-275} \\ 25 \end{array} = 5.5\%$$

ANSWERS TO SELF-TEST 14.3

1. 400%

2. 100%

3. 150%

4. 0

1. | South Asian | 1.9 million |
    | East Asian | 1.8 million |
    | Black | 1.2 million |
    | Other | 2.7 million |
    | Total | 7.6 million |

    South Asian: $\dfrac{1.9}{7.6} = 25.0\%$ *Check:* 25.0%

    East Asian: $\dfrac{1.8}{7.6} = 23.7\%$ 23.7

    Black: $\dfrac{1.2}{7.6} = 15.8\%$ 15.8

    Other: $\dfrac{2.7}{7.6} = 35.5\%$ 35.5

    100.0%

2. | freshmen | 700 |
    | sophomores | 650 |
    | juniors | 600 |
    | seniors | 550 |
    | | 2,500 |

    freshmen $= \dfrac{700}{2,500} = \dfrac{7}{25} = 28\%$

    sophomores $= \dfrac{650}{2,500} = \dfrac{65}{250} = \dfrac{13}{50} = 26\%$ *Check:* 28%

    juniors $= \dfrac{600}{2,500} = \dfrac{60}{250} = \dfrac{12}{50} = 24\%$ 26

    seniors $= \dfrac{550}{2,500} = \dfrac{55}{250} = \dfrac{11}{59} = 22\%$ 24

    22

    100%

3. | consumers | $200 million |
    | businesses | 300 million |
    | government | 500 million |
    | | $1,000 million |

    consumers $= \dfrac{200}{1,000} = 20\%$

    businesses $= \dfrac{300}{1,000} = 30\%$ *Check:* 20%

    government $= \dfrac{500}{1,000} = 50\%$ 30

    50

    100%

1. 10% of $20 is $2, and one-half of $2 is $1. So you would leave a $3 tip ($2 + $1).

2. Leave 15% of $30, which comes to $4.50. (10% of $30 is $3. Half of $3 is $1.50: $3 + $1.50 = $4.50.)

3. 10% of $45 is $4.50. Double $4.50 and leave a tip of $9.00

4. Round $73.95 to $74. 10% of $74 is $7.40. Double $7.40 to $14.80. Your tip would be $14.80.

15 Solving Simple Equations

Throughout this chapter, we'll be talking about equations. What's an equation? It's a mathematical statement that tells us that what is on one side of an equal sign (=) is equal to what's on the other side. For example, $3 + 4 = 7$. The equations that need solving are ones that include a letter or symbol, called a variable, that stands for some unknown number. Solving is figuring out what the unknown is. In algebra, it's common to use letters like x, y, and z as variables, but variables aren't really new to you. If you can fill in the box in the problem below, you've solved for a variable.

$$\Box + 1 = 2$$

This chapter will ask you to use ideas from Chapter 2, "Essential Arithmetic," Chapter 5, "Mental Math," Chapter 7, "Fractions," and Chapter 8, "Decimals." Review before you begin, or refer back to those chapters as you work through.

1 THE CARE AND TREATMENT OF EQUATIONS

Now suppose we decide to add 3 to the left side of the equation $3 + 4 = 7$ that we mentioned above: $3 + 4 + 3$. What must we do to keep the equation in balance? That's right! We must add 3 to the right side: $7 + 3$. Think of an equation like a balance scale, the type with two pans that you can place things in. If the pans are balanced, and you add something to one pan, it will sink down, and you'll have to add the same amount to the other to get them back in balance. If $3 + 4 = 7$, then $3 + 4 + 3 = 7 + 3$.

Let's try another equation: $5 + 4 = 9$. If we subtracted 2 from the left side, $5 + 4 - 2$, what other change is needed? You guessed it! We must subtract 2 from the right side: $9 - 2$.

It follows that if the right side were multiplied by a certain number, we'd have to multiply the left side by that same number. Finally, if one side is divided by 8, for example, the other side must also be divided by 8.

So what we do to one side of the equation, we must do to the other side. Why? Because the two sides must always equal.

2 ISOLATING *X*

In algebra, x is often used to designate the unknown quantity—hence, the quantity we are seeking to find. You'll see other letters used as well, but x is the most common. Since we're trying to determine the value of this unknown quantity, it would be helpful to have it sitting all alone on one side of the equal sign and a number on the other. There's no mystery to the equation $x = 7$. You might be able to guess the value of x if the equation is $x + 8 = 10$, but the more complicated the equation becomes, the harder it is to figure out the value of the unknown by just looking. The ideas above, all about doing the same thing to both sides of an equation, will go a long way toward isolating the unknown.

3 ADDITION AND SUBTRACTION WITH *X*

We can write $x +$ a number or $x -$ a number but we can't actually do that arithmetic because we don't know what x is. Our goal, instead, is to eliminate the arithmetic by doing the opposite operation on both sides of the equation. That will keep the equation in balance and leave x all alone on its side.

Problem 1:

If $x - 6 = 2$, how much is x?

Solution:

$$x - 6 = 2$$
$$x - 6 + 6 = 2 + 6$$
$$x = 8$$

To isolate x, we added 6 to both sides of the equation. Why did we add 6? Why not 12? Why not 128? Because adding 6 will "undo" subtracting 6 and leave x all alone.

Now let's check our work. Go back to the original equation, $x - 6 = 2$ and substitute your answer, 8, for x. What do you get?

$$x - 6 = 2$$
$$8 - 6 = 2$$
$$2 = 2$$

So it checks.

Problem 2:

If $x - 4 = 9$, how much is x?

Solution:

$$x - 4 = 9$$
$$x - 4 + 4 = 9 + 4$$
$$x = 13$$

Here we added 4 to both sides.

Check your work here, by substituting your answer into the original equation. Does it check out? We'll do it once more:

$$x - 4 = 9$$
$$13 - 4 = 9$$
$$9 = 9$$

Problem 3:

If $x + 3 = 4$, how much is x?

Solution:

$$x + 3 = 4$$
$$x + 3 - 3 = 4 - 3$$
$$x = 1$$

Here we subtracted 3 from both sides to isolate x.

Problem 4:

If $x + 6 = 9$, how much is x?

Solution:

$$x + 6 = 9$$
$$x + 6 - 6 = 9 - 6$$
$$x = 3$$

Still a little confused about whether to add or subtract? Then remember, we want to isolate x. In the first problem, $x - 6 = 2$, we had to get rid of the -6. We did that by adding $+6$ to both sides. In each of the subsequent problems, we isolated x by getting rid of the number that was on x's side of the equation. And what we did on one side of the equation, we did on the other side as well.

SELF-TEST 15.1

Find x in each of these problems:

1. $x - 8 = 17$ 2. $x - 4 = 5$ 3. $x + 4 = 12$ 4. $x + 10 = 15$

---

4   MULTIPLICATION AND DIVISION WITH *X*

When the reason x is not alone is that something is added to it, we subtracted from both sides. When we saw x minus a number, we added to both sides. So, when the reason x is not all alone is that it's multiplied by a number, we'll divide both sides by that number. If x is divided by a number, multiplying both sides of the equation by that number will isolate x.

Problem 1:
We'll begin with an equation: $2x = 50$. How much is x?

Solution:

$$2x = 50$$

$$x = 25$$

How do we know that x is 25? You might see it just from your experience. If you have 2 of the same kind of coin, and they're worth a total of 50 cents, you can probably figure out the coins are quarters, each worth 25 cents. When in doubt, we divide both sides of the equation by 2 to isolate x.

Run your check on the answer by substituting it into the original equation.

Check:

$$2x = 50$$

$$2 \times 25 = 50$$

$$50 = 50$$

Here's another one.

Problem 2:

If $3x = 12$, how much is x?

Solution:

$$3x = 12$$

$$x = \frac{12}{3}$$

$$x = 4$$

Problem 3:

Find x when $\frac{x}{3} = 2$.

Solution:

$$\frac{x}{3} = 2$$

$$\frac{\cancel{3}}{1} \times \frac{x}{\cancel{3}} = 3 \times 2$$

$$x = 6$$

We multiplied both sides of the equation by 3.

Problem 4:

Solve for x when $\frac{x}{4} = 10$.

Solution:

$$\frac{x}{4} = 10$$

$$\frac{\cancel{4}}{1} \frac{x}{\cancel{4}} = 4 \times 10$$

$$x = 40$$

SELF-TEST 15.2

Find x in each of these problems:

1. $4x = 8$ 2. $6x = 15$ 3. $\frac{x}{2} = 9$ 4. $\frac{x}{5} = 4$

5   FRACTIONS AND DECIMALS IN EQUATIONS

All those whole numbers are nice, but you know that life isn't always that simple. You'll need to solve some equations that have fractions or decimals, and certainly some that produce answers that are fractions or decimals. You're still after just one thing—you want to get x all alone. And you'll use the same strategy: do the opposite of the operation you see, and do it to both sides. Don't spend a lot of time converting fractions and decimals. If the equation starts with decimals, stay with decimals. If it starts with fractions, stay with fractions.

Problem 1:

If $0.2x = 8$, how much is x?

Solution:

$$0.2x = 8$$
$$\frac{0.2x}{0.2} = \frac{8}{0.2}$$
$$x = 40$$

To isolate x, we divided both sides by 0.2. What we do to one side of the equation, we must do to the other side.

Does this answer check out? There's only one way to find out.

Check:

$$0.2x = 8$$
$$0.2 \times 40 = 8$$
$$8 = 8$$

Problem 2:

If $2.5x = 17$, how much is x?

Solution:

$$2.5x = 17$$
$$\frac{2.5x}{2.5} = \frac{17}{2.5} = \frac{170}{25}$$
$$x = 6.8$$

Check:

$$
\begin{array}{r}
6.8 \\
\times 2.5 \\
\hline
3\ 40 \\
13\ 6 \\
\hline
17.00
\end{array}
$$

We'll work out one more.

Problem 3:

Find x when $2\frac{1}{2}x = 10$.

Solution:

$$2\frac{1}{2}x = 10$$

$$\frac{5}{2}x = 10$$

$$\frac{5}{2}x \div \frac{5}{2} = 10 \div \frac{5}{2}$$

$$\frac{\cancel{5}}{\cancel{2}} \times \frac{\cancel{2}}{\cancel{5}}x = \frac{\cancel{10}^{2}}{1} \times \frac{2}{\cancel{5}}$$

$$x = 4$$

SELF-TEST 15.3

Find x in each of these problems. Round to the nearest tenth when necessary.

1. $0.4x = 9$ 2. $\frac{9}{10}x = 12$ 3. $1\frac{4}{5}x = 2$ 4. $3.5x = 15$

6   NEGATIVE NUMBERS

Can x ever be a negative number? It sure can. Suppose that x represents your losses at the track this year. Or the money you spent on lottery tickets.

Problem 1:

If $x + 7 = 3$, how much is x?

Solution:

$$x + 7 = 3$$
$$x + 7 - 7 = 3 - 7$$
$$x = -4$$

What we do here is follow the same procedure we used in the first section of this chapter to isolate x. If you're at all uncertain about this procedure, please return to Frame 3.

Problem 2:

If $x + 10 = 2$, how much is x?

Solution:

$$x + 10 = 2$$
$$x + 10 - 10 = 2 - 10$$
$$x = -8$$

Problem 3:

Find x if $x + 15 = -4$.

Solution:

$$x + 15 = -4$$
$$x + 15 - 15 = -4 - 15$$
$$x = -19$$

We hope you're still checking your answers.

Check:

$$x + 15 = -4$$
$$-19 + 15 = -4$$
$$-4 = -4$$

Problem 4:

Find x if $x - 2 = -5$.

Solution:

$$x - 2 = -5$$
$$x - 2 + 2 = -5 + 2$$
$$x = -3$$

Problem 5:

Find x when $x - 6 = -7$.

Solution:

$$x - 6 = -7$$
$$x - 6 + 6 = -7 + 6$$
$$x = -1$$

If negative numbers make you nervous, then you need to review Chapter 6. After you've done that, return to Frame 6.

If you did not have any difficulty with these problems, then you shouldn't have any difficulty with those in Self-Test 15.4.

SELF-TEST 15.4

Find x in each of these problems:

1. $x + 4 = -8$ 2. $x + 9 = 2$ 3. $x - 7 = -10$
4. $x + 5 = 4$ 5. $x + 3 = 2$ 6. $x - 10 = -14$

7   MULTI-STEP EQUATIONS

The problems in this section combine the various things we've covered in this chapter—addition, subtraction, multiplication, division, decimals, and negative numbers. Perform the opposite operation in the opposite order. If the equation was formed by first multiplying x by 7 and then adding 3, you should subtract 3 and then divide by 7.

Problem 1:

If $2x + 5 = 10$, solve for x.

Solution:

$$2x + 5 = 10$$

$$2x = 5$$

$$x = \frac{5}{2}$$

$$x = 2\frac{1}{2}$$

What we've done here is combine two operations to isolate x. First we got rid of the 5 by subtracting it from both sides of the equation. This left us with $2x$. To get x, we divided both sides by 2.

Problem 2:

If $5x - 3 = 9$, solve for x.

Solution:

$$5x - 3 = 9$$

$$5x = 12$$

$$x = \frac{12}{5}$$

$$x = 2\frac{2}{5}$$

Problem 3:

Find x when $\frac{x}{3} + 2 = 5$.

Solution:

$$\frac{x}{3} + 2 = 5$$

$$\frac{x}{3} = 3$$

$$x = 9$$

Problem 4:

How much is x when $\frac{2}{3}x - 1 = 8$?

Solution:

$$\frac{2}{3}x - 1 = 8$$

$$\frac{2}{3}x = 9$$

$$\frac{3}{2} \times \frac{2}{3}x = \frac{3}{2} \times \frac{9}{1}$$

$$x = \frac{27}{2}$$

$$x = 13\frac{1}{2}$$

Problem 5:

Find x when $6x + 11 = 3$.

Solution:

$$6x + 11 = 3$$

$$6x = -8$$

$$x = -1\frac{1}{3}$$

Problem 6:

When $\frac{3}{5}x - 2 = -10$, how much is x?

Solution:

$$\frac{3}{5}x - 2 = -10$$

$$\frac{3}{5}x = -8$$

$$\frac{5}{3} \times \frac{3}{5}x = \frac{5}{3} \times \frac{-8}{1}$$

$$x = \frac{-40}{3}$$

$$x = -13\frac{1}{3}$$

Problem 7:

If $0.5x - 1 = -6$, find x.

Solution:

$$0.5x - 1 = -6$$
$$0.5x = -5$$
$$x = -10$$

Problem 8:

When $2.8x + 12 = 3$, how much is x?

Solution:

$$2.8x + 12 = 3$$
$$2.8x = -9$$
$$\frac{2.8x}{2.8} = -\frac{9}{2.8} = -\frac{90}{28}$$
$$x \approx -3.2$$

SELF-TEST 15.5

Find x in each of these problems:

1. $4x + 7 = 15$ 2. $2x - 7 = 13$ 3. $\frac{x}{5} + 4 = 19$ 4. $\frac{3}{4}x - 1 = -6$

5. $9x + 1 = 0$ 6. $0.7x + 3 = 15$ 7. $0.4x + 12 = 3$ 8. $3.1x - 9 = -1$

ANSWERS TO SELF-TEST 15.1

1. $x = 25$ 2. $x = 9$ 3. $x = 8$ 4. $x = 5$

ANSWERS TO SELF-TEST 15.2

1. $x = 2$ 2. $x = 2\frac{1}{2}$ 3. $x = 18$ 4. $x = 20$

ANSWERS TO SELF-TEST 15.3

1. $\dfrac{0.4x}{0.4} = \dfrac{9}{0.4} = \dfrac{90}{4}$

    $x = 22.5$

2. $\dfrac{9}{10}x = 12$

$\dfrac{9}{10}x \div \dfrac{9}{10} = 12 \div \dfrac{9}{10} = \dfrac{\overset{4}{\cancel{12}}}{1} \times \dfrac{10}{\underset{3}{\cancel{9}}}$

            $x = \dfrac{40}{3} = 13\dfrac{1}{3}$

3. $1\dfrac{4}{5}x = 2$

$\dfrac{9}{5}x \div \dfrac{9}{5} = \dfrac{2}{1} \div \dfrac{9}{5} = \dfrac{2}{1} \times \dfrac{5}{9}$

    $x = \dfrac{10}{9} = 1\dfrac{1}{9}$

4. $3.5x = 15$

   $\dfrac{3.5x}{3.5} = \dfrac{15}{3.5} = \dfrac{150}{35}$

     $x \approx 4.3$

ANSWERS TO SELF-TEST 15.4

1. $x = -12$ 2. $x = -7$ 3. $x = -3$

4. $x = -1$ 5. $x = -1$ 6. $x = -4$

ANSWERS TO SELF-TEST 15.5

1. $4x + 7 = 15$
        $4x = 8$
         $x = 2$

2. $2x - 7 = 13$
       $2x = 20$
        $x = 10$

3. $\dfrac{x}{5} + 4 = 19$
      $\dfrac{x}{5} = 15$
       $x = 75$

4. $\dfrac{3}{4}x - 1 = -6$
      $\dfrac{3}{4}x = -5$
     $3x = -20$
      $x = -6\dfrac{2}{3}$

5. $9x + 1 = 0$
      $9x = -1$
      $x = -\dfrac{1}{9}$

6. $0.7x + 3 = 15$
     $0.7x = 12$
      $x = \dfrac{12}{0.7} = \dfrac{120}{7}$
      $x \approx 17.1$

7. $0.4x + 12 = 3$
     $0.4x = -9$
   $x = -\dfrac{9}{0.4} = -\dfrac{90}{4}$
      $x = -22.5$

8. $3.1x - 9 = -1$
     $3.1x = 8$
   $x = \dfrac{8}{3.1} = \dfrac{80}{31}$
      $x \approx 2.6$

16 Powers and Roots

When we use the same number several times in a multiplication, it's convenient to have a shorthand to express it. Another way of writing $2 \times 2 \times 2 \times 2$ is 2^4, or 2 to the fourth power. It's a time-saver for a small number of repeats, but nearly essential if, say, you wanted to multiply 10 or 20 copies of a number. (In algebra, this will be extremely important, but more on that later.)

You may have noticed in past chapters that we often ask you to work backwards, and we will do that again here, in a process called finding a root. It's a bit complicated when you're talking powers and roots, and we'll focus on estimation for much of our conversation on roots. Calculators will be helpful for finding both large powers and for finding roots, but as always, you want to understand enough to know if your calculator's answer is reasonable.

In this chapter, we'll make reference back to Chapter 2, "Essential Arithmetic," Chapter 7, "Fractions," and Chapter 5, "Mental Math."

1 EXPONENTS

What is an exponent? An *exponent* is a small number, written raised and to the right of another, that tells you how many copies of that number are used in the multiplication. The number that's actually being multiplied is called the *base*, and the small number that tells you how many copies of that base to multiply is the *exponent*. Together, the base and the exponent make a *power*.

In algebra, you may see a variable, like x, used as the base of a power. The second power is a number times itself: $3^2 = 3 \times 3$; $5^2 = 5 \times 5$. And x^2? That's x times x. Finding the second power is often called squaring, and 7^2 may be read "7 squared." (This comes from the area of a square.) The third power is a number times itself times itself again. So $3^3 = 3 \times 3 \times 3$; $5^3 = 5 \times 5 \times 5$. And $x^3 = x$ times x times x.

The terms x^2 and x^3 are powers of some unknown number x. The exponents are the little numbers to the right and just above each x. The first term, x^2, is commonly read x squared; the other, x^3, can be read x to the third or x cubed. The second and third powers are the ones you'll see most often, so they're usually just called squares and cubes.

Problem 1:

How much is x^2 if $x = 4$?

Solution:

$$x^2 = 4^2 = 4 \times 4 = 16$$

Problem 2:

How much is x^2 if $x = 6$?

Solution:

$$x^2 = 6^2 = 6 \times 6 = 36$$

Next, we'll do cubes.

Problem 3:

How much is x^3 if $x = 2$?

Solution:

$$x^3 = 2^3 = 2 \times 2 \times 2 = 8$$

Problem 4:

How much is x^3 if $x = 3$?

Solution:

$$x^3 = 3^3 = 3 \times 3 \times 3 = 27$$

Now, while we're hot, we might as well talk about x^1, or x to the first power. That's the same as x. You're using one x. For example, if x is 7, how much is x^1? The answer is 7. And if $x^1 = 4$, how much is x? It is 4.

It's possible to have exponents like zero, or negative numbers, or even fractions, but what they mean and how they came to mean that takes more explanation than we can handle right now. Just so you have working information, we'll give you these rules:

1. Any number except zero, to the zero power, equals one.

2. If you see a negative exponent, like 7^{-3}, just rewrite it as one over the same base with a positive exponent: $\frac{1}{7^3}$.

3. If you see a fractional exponent, the numerator tells the power and the denominator tells the root. (More on roots in Frame 2.)

SELF-TEST 16.1

1. If $x = 3$, how much is x^4?

2. If $x = 5$, how much is x^2?

3. If $x = 2$, how much is x^3?

4. If $x = 8$, how much is x^1?

5. If $x = 10$, how much is x^0?

6. If $x = 3$, how much is x^{-2}?

7. If $x = 4$, how much is

   (a) x^0

   (b) x^1

   (c) x^2

   (d) x^3

   (e) x^4

The Powers That Be

The easiest number to raise to a power is 1, because $1 \times 1 = 1$, so 1 to any power is 1. And here's a summary of the powers of 10:

$$10^0 = 1$$
$$10^1 = 10$$
$$10^2 = 100$$
$$10^3 = 1,000$$
$$10^4 = 10,000$$

Can you see the pattern? Count the zeros.

---

2 ROOTS

What's a root? Not the kind you find in the garden but the mathematical kind. Suppose you know a number, like 25, can be thought of as some number to a power. In this case, $25 = 5^2$. The square root of 25 is the number it grew from, the 5. The square root of a number is the number you would square to produce it.

What's the square root of 4? What number times itself gives us 4? The answer could be 2. So 2 is a square root of 4. But numbers can have two square roots. It's true that $2 \times 2 = 4$, making 2 a square root of 4, but it's also true that $-2 \times -2 = 4$, so -2 is also a square root of 4. When we want to be thorough and give both possible square roots, we can use the $\pm$ sign. The square root of 4 is ± 2.

We denote square root by the sign $\sqrt{}$, called a radical, from the Latin word for root. For example, $\sqrt{4}$ means the principal, or positive, square root of 4. If we want the negative square root, we write $-\sqrt{4}$ and if we want both, $\pm\sqrt{4}$.

Problem 1:

Find $\sqrt{9}$. What number times itself is 9?

Solution:

The answer could be 3 or -3. The square root of 9 is ± 3. $\sqrt{9} = 3$, $-\sqrt{9} = -3$, and $\pm\sqrt{9} = \pm 3$.

Now we'll do another one.

Problem 2:

How much is the square root of 16?

Solution:
That's right. It's ± 4.

Figure this one out:

Problem 3:

If $x^2 = 36$, how much is x?

Solution:
The answer is ± 6.

In some problems, you'll be able to see that only the positive square root makes sense. You can't draw a square -6 feet long. In cases that aren't clear, give both the positive and the negative square root.

You may have noticed that we didn't try to figure out the square roots of just any old numbers. The numbers we chose—4, 9, 16, and 36—all had nice whole number square roots. What about other numbers, such as 5, 6, 7, or 8? These numbers all have squares roots of more than 2, but less than 3, or more than -3 but less than -2, if we consider the negative square roots. Let's focus on the principal, or positive, square roots for now. The square root of 6 must be bigger than 2, because $2^2 = 4$, but it can't be 3 because $3^2 = 9$. It's a decimal, one of the ones we mentioned in Chapter 9, "Converting Fractions and Decimals," when we were discussing decimals that go on forever but don't repeat.

And how would we find the square roots like these? The chances are, you'll never have to. Some texts (not this one) print tables of square roots. Some math texts (not this one) show you how to figure out square roots. There's an old pencil and paper method that sort of looks like long division, but isn't, and there's the trial-and-error method, which is, well, trying. This is a place where it makes sense to either write $\sqrt{6}$ and be done with it or take advantage of a calculator or computer. Any decimal a calculator gives you will be rounded so it's an approximate value of the square root. Most of the time, that's fine. If you want to say the exact square root of six, write $\sqrt{6}$.

SELF-TEST 16.2

Find the square root of

1. 100　　　　　2. 64　　　　　3. 49　　　　　4. 81

How much is x if

5. $x^2 = 9$　　　　6. $x^2 = 25$　　　　7. $x^2 = 16$　　　　8. $x^2 = 4$

3 ADVANCED PROBLEMS

Now that you're comfortable solving equations that contain x^2, let's include some other operations. The job is still to get x by itself, and to do that you'll need to first get x^2 by itself. As always, undo addition by subtracting, and undo multiplying by dividing.

Problem 1:

If $x^2 + 4 = 20$, find x.

Solution:

$$x^2 + 4 = 20$$
$$x^2 = 16$$
$$x = \pm 4$$

Problem 2:

If $2x^2 = 50$, find x.

Solution:

$$2x^2 = 50$$
$$x^2 = 25$$
$$x = \pm 5$$

Problem 3:

If $3x^2 - 5 = 7$, find x.

Solution:

$$3x^2 - 5 = 7$$
$$3x^2 = 12$$
$$x^2 = 4$$
$$x = \pm 2$$

Problem 4:

If $x = 10$, how much is $x^2 - 5$?

Solution:

$$x^2 - 5 =$$
$$10^2 - 5 =$$
$$100 - 5 = 95$$

Problem 5:

If $x = 3$, how much is $2x^2 + 1$?

Solution:

$$2x^2 + 1 =$$
$$2 \times 3^2 + 1 =$$
$$2 \times 9 + 1 =$$
$$18 + 1 = 19$$

Problem 6:

If $x = 8$, how much is $3x^2 + 2x - 3$?

Solution:

$$3x^2 + 2x - 3 =$$
$$(3 \times 8 \times 8) + (2 \times 8) - 3 =$$
$$192 + 16 - 3 = 205$$

SELF-TEST 16.3

How much is x?

1. $2x^2 - 90 = 72$

2. $10 + 3x^2 = 37$

3. $5x^2 + 1 = 126$

4. $7 - 4x^2 = -57$

Solve each of these problems:

5. If $x = 8$, how much is $150 - 2x^2 + x$?

6. If $x = 1$, how much is $2x^2 - 1$?

7. If $x = 7$, how much is $14 - x^2$?

8. If $x = 10$, how much is $3x^2 + 4$?

ANSWERS TO SELF-TEST 16.1

1. $3 \times 3 \times 3 \times 3 = 81$ 2. $5 \times 5 = 25$ 3. $2 \times 2 \times 2 = 8$

4. 8 5. 1 6. $\frac{1}{3^2} = \frac{1}{3 \times 3} = \frac{1}{9}$

7. (a) 1

   (b) 4

   (c) $4 \times 4 = 16$

   (d) $4 \times 4 \times 4 = 64$

   (e) $4 \times 4 \times 4 \times 4 = 256$

ANSWERS TO SELF-TEST 16.2

1. ±10
2. ±8
3. ±7
4. ±9
5. ±3
6. ±5
7. ±4
8. ±2

ANSWERS TO SELF-TEST 16.3

1. $2x^2 = 162$
   $x^2 = 81$
   $x = \pm9$

2. $3x^2 = 27$
   $x^2 = 9$
   $x = \pm3$

3. $5x^2 = 125$
   $x^2 = 25$
   $x = \pm5$

4. $-4x^2 = -64$
   $4x^2 = 64$
   $x^2 = 16$
   $x = \pm4$

5. $150 - (2 \times 8 \times 8) + 8 = 150 - 128 + 8 = 30$

6. $(2 \times 1 \times 1) - 1 = 2 - 1 = 1$

7. $14 - (7 \times 7) = 14 - 49 = -35$

8. $(3 \times 10 \times 10) + 4 = 300 + 4 = 304$

17 Very Large and Very Small Numbers

The main purpose of this book is to put you at ease with numbers. So far, we've concentrated on relatively small numbers—fractions, decimals, hundreds of dollars, thousands of dollars.

Well, how do you feel about *very* large numbers? Billions? Trillions? Quadrillions? Quintillions? Can you divide four hundred thirty-two quadrillion by one hundred and eight trillion? Did you get four thousand? If you did, then you definitely know your big numbers and can move all the way around the board, pass GO, and collect $200. But hang around for the parts of this chapter that will talk about convenient ways to write and work with very large and very small numbers.

We'll be using ideas from Chapter 16, "Powers and Roots," Chapter 7, "Fractions," Chapter 8, "Decimals," and Chapter 5, "Mental Math." Look back to those if you run into trouble.

1 WRITING LARGE NUMBERS

If you won a million dollars in the lottery, you would know how to count it, but have you ever thought about how long it would take? Of course, that would depend on how you were paid your winnings. Hundred-dollar bills? Ten-dollar bills? One-dollar bills? Pennies? If you received one million one-dollar bills, and counted one per second—with no breaks, at all, for any reason—it would take 11 days, 13 hours, 46 minutes, and 40 seconds to count from one to one million.

Problem 1:
Write out the number one million—with all its zeros.

Solution:

$$1,000,000$$

Remember the quick way of doing powers of ten? One million is a power of ten. It's 10^6. When numbers get really large, it can help to use powers where we can.

Problem 2:
If, instead of one million one-dollar bills, you were paid with ten-dollar bills, how many bills would you have to count?

Solution:

1,000,000 ÷ 10 = 100,000 ten-dollar bills. That would only take 1 day, 3 hours, 46 minutes, and 40 seconds to count one ten-dollar bill every second.

Did you remember that dividing by 10 cancels one zero? If you write one million as 10^6, you can think of 10 as 10^1, and $10^6 ÷ 10^1 = 100,000 = 10^5$. Do you see the shortcut here? When you divide powers of 10, you keep the 10 and subtract the exponents. It can make arithmetic with large numbers much easier. (This actually works for any number. $7^5 ÷ 7^2 = 7^3$. $16^{10} ÷ 16^8 = 16^2$. But it's most helpful with powers of 10.)

Problem 3:

If, instead of one million one-dollar bills, you were paid in pennies (eek!), how many pennies would you have to count?

Solution:

There are 100 pennies for each dollar, so you'd have to count $100 × 1,000,000 = 100,000,000$ pennies. If you write 100 as 10^2 and 1,000,000 as 10^6, that's $10^2 × 10^6 = 100,000,000 = 10^8$. There's a helpful shortcut here too. To multiply powers of ten, keep the ten and add the exponents. By the way, those pennies would keep you counting a penny a second for 3 years, 62 days, 9 hours, 46 minutes, and 40 seconds.

Our chances of winning the lottery are slim, but let's look at some large numbers you might encounter. Our number system is based on ten, so we have a place for ones, then to the left of the ones are the tens, then 10^2 or hundreds to the left of the tens. Numbers larger than three digits get separated into groups of three digits by commas. The thousands group (or thousands period) from right to left include thousands (10^3), ten-thousands (10^4), and hundred-thousands (10^5).

Problem 4:

Write out the number one hundred fifty-three thousand.

Solution:

153,000

One hundred fifty-three thousand is the number one hundred fifty-three followed by a comma and three zeros.

The next group starts with the millions (10^6). One million is designated by a one, a comma, three zeros, another comma, and three more zeros. One million is a one and two sets of three zeros. The millions group includes millions, ten-millions, and hundred-millions.

Problem 5:

Please write the number eight hundred four million.

Solution:

$$804,000,000$$

Problem 6:

Please write the number two hundred sixteen million, four hundred seventy-six thousand.

Solution:

$$216,476,000$$

Now we'll reverse the process.

Problem 7:

Please express this number in words: 547,302.

Solution:

Five hundred forty-seven thousand, three hundred two.

Problem 8:

Express 81,963,000 in words.

Solution:

Eighty-one million, nine hundred sixty-three thousand.

Problem 9:

Express this number in words: 905,100,515.

Solution:

Nine hundred five million, one hundred thousand, five hundred fifteen

SELF-TEST 17.1

Please translate these words into numbers:

1. Four hundred seventy-five thousand, five hundred

2. Two million

3. Seven hundred seventy-four million, two hundred fifty thousand

4. Ninety-three thousand two

5. Three hundred forty-six million, five hundred sixty-one thousand, two hundred eighty-eight

Please express these numbers in words:

6. 75,000   7. 400,160,000   8. 145,005

9. 785,631,072   10. 100,175,200

2   SCIENTIFIC NOTATION

When you see a large number like the 804,000,000 from Problem 5 in Frame 1, it may be written as 804 million so that you don't have to count zeros to understand it, and because printing all the digits of very large numbers takes up a lot of space. In a scientific or technical document, you may see 804,000,000 in the form 8.04×10^8. This is called *scientific notation*.

Scientific notation is a method of writing a large number in a compact form. In scientific notation, a large number is written as a number greater than or equal to 1 but less than 10, times a power of 10. It's possible to write a large number like 804 million in many different ways. For example, 804,000,000 = 804 × 1,000,000 or 804 ×10^6. But it could also be 80.4 ×10^7 or 8.04×10^8. Scientific notation agrees to use that last form.

To change 804,000,000 to scientific notation:

• Count the number of digits between the first digit and the location where the decimal point would be to determine the exponent. In this number, that's eight places. 804,000,000

          8 places

• Place a decimal point after the first digit, in this case, 8. You can drop all the zeros after the 4.

• Multiply 8.04 by 10^8. The exponent of 8 came from the count of digits.

$$804,000,000 = 8.04 \times 10^8$$

Problem 1:

Write 325,000,000 in scientific notation.

Solution:

325,000,000 has eight digits between the 3 and the decimal point. 325,000,000 $= 3.25 \times 10^8$.

Problem 2:

Write 47,300,000 in scientific notation.

Solution:

There are seven digits between the 4 and the decimal point. 47,300,000 $= 4.73 \times 10^7$.

To change a number in scientific notation to standard form:

- Note the exponent on the 10. This is the number of places you will move the decimal point. The number 4.59×10^6 has an exponent of 6, so you will need to move the decimal point six places.
- If there are not enough digits after the decimal point, add zeros. $4.59 \times 10^6 = 4.590000 \times 10^6$.
- Move the decimal point, place commas where needed, and eliminate the power of ten.
$$4.59 \times 10^6 = 4.590000 \times 10^6 = 4,590,000$$

Problem 3:

Write 6.1×10^9 in standard form.

Solution:

$$6.1 \times 10^9 = 6.100000000 \times 10^9 = 6,100,000,000$$

SELF-TEST 17.2

Write these numbers in scientific notation.

1. 275,600,000   2. 9,240,000   3. 63,400,000

4. 58,000   5. 30,042,500

Write these numbers in standard notation.

6. 6.27×10^5 7. 5.84×10^8 8. 7.1×10^6

9. 9×10^5 10. 3.01×10^{11}

3 MILLIONS, BILLIONS, TRILLIONS, AND BEYOND

You might win a million dollars in the lottery, but it's highly unlikely that you'll ever win a billion. Numbers in the billions and trillions rarely turn up in anything but government budgets, astronomers' measurements, or terrifying predictions. But you should be able to read and write them, even if it's difficult to wrap your mind around them.

Millions, Billions, Trillions, and Beyond		
one million	1,000,000	10^6
one billion	1,000,000,000	10^9
one trillion	1,000,000,000,000	10^{12}
one quadrillion	1,000,000,000,000,000	10^{15}
one quintillion	1,000,000,000,000,000,000	10^{18}

Problem 1:

Write out the number seven hundred billion, three hundred and fifty-five million in standard form and in scientific notation.

Solution:

$$700,355,000,000 \text{ or } 7.00355 \times 10^{11}$$

You'll often see numbers this large rounded to make them easier to understand. For example, 7.00355×10^{11} might be expressed as 7×10^{11}.

Problem 2:

Write out the number four hundred and three billion, two hundred nineteen million, seven hundred forty-two thousand, and one in standard form and in scientific notation.

Solution:

$403,219,742,001$ or $4.03219742001 \times 10^{11}$, but you'd most likely see it as 4.03×10^{11}.

Problem 3:

How would you write one trillion, in standard form and in scientific notation?

Solution:

$$1,000,000,000,000 \text{ or } 1 \times 10^{12}$$

Problem 4:

In 2020, the U.S. Gross Domestic Product, or GDP, was $20.94 × 10^{12}$. Write that number in standard form. Can you write it in words?

Solution:

Our 2020 GDP was $20,940,000,000,000. In other words, twenty trillion, nine hundred forty billion dollars.

Problem 6:

Express 7,500,000,000 in words.

Solution:

Seven billion, five hundred million

Problem 7:

Now express 511,388,950,000,000 in words.

Solution:

Five hundred eleven trillion, three hundred eighty-eight billion, nine hundred fifty million.

How's it going? If none of this throws you, do Self-Test 17.3. But if you're still not comfortable with millions, billions, and trillions, go back to Frame 1. It's often better the second time around.

SELF-TEST 17.3

Please translate these words into numbers, in standard form and in scientific notation:

1. fifty billion

2. four hundred trillion, two hundred and thirty billion

3. seventy-two trillion, nine hundred fifty-four billion, three hundred and eight million

4. ten trillion

5. five hundred and two trillion, four hundred sixty-four billion, nine hundred and seventeen million

Please express these numbers in words:

6. 1.75×10^{11} 7. 8×10^{10} 8. 675,505,768,400,000

9. 3×10^{12} 10. 234,459,875,361,827

4   MULTIPLYING LARGE NUMBERS

How much is $1,000 \times 1,000$? When we multiply any whole number (i.e., not a fraction or decimal) by 1,000, we just add three zeros, so $1,000 \times 1,000 = 1,000,000$. One million is one thousand thousands.

How much is $1,000 \times \$1,000,000$? That's one thousand millions or one million thousands. Either way, it comes to $\$1,000,000,000$, or one billion dollars.

And finally, what's $1,000 \times 1,000,000,000$? It's $1,000,000,000,000$, or one trillion. So one thousand billion is one trillion.

But the multiplying we need to do won't always have just ones and zero. Now we'll deal with numbers that don't all end in zero.

Problem 1:
How much is $10,000 \times 1,475$?

Solution:

$$14,750,000$$

Problem 2:
How much is $100,000 \times 1,302,116$?

Solution:

$$130,211,600,000$$

When we multiply by 10, we add one zero; when we multiply by 100, we add two zeros; when we multiply by 1,000, we add three zeros, and so forth. Don't bother to memorize all these rules. Just multiply the digits that aren't trailing zeros, then count up the total number of zeros and tack them on.

Problem 3:
How much is $16,000 \times 4,000$?

Solution:
Think of this as $16 \times 1,000 \times 4 \times 1,000 = 16 \times 4 \times 1,000 \times 1,000 = 64 \times 1,000,000 = 64,000,000$

Problem 4:

How much is $134,100 \times 200,000$?

Solution:

$$1,341 \times 2 \times 100 \times 100,000 = 2,682 \times 10,000,000 = 26,820,000,000$$

One reason people who work with big numbers like scientific notation is because it simplifies multiplication and division. Let's look at those same problems again, using scientific notation. Multiply the numbers that are not powers of ten, then multiply the powers of ten by keeping the ten and adding the exponents.

Problem 5:

How much is $1.6 \times 10^4 \times 4 \times 10^3$?

Solution:

$$1.6 \times 10^4 \times 4 \times 10^3 = 1.6 \times 4 \times 10^4 \times 10^3 = 6.4 \times 10^{4+3} = 6.4 \times 10^7$$

$$= 64,000,000$$

Problem 6:

How much is $1.341 \times 10^5 \times 2 \times 10^5$?

Solution:

$$1.341 \times 10^5 \times 2 \times 10^5 = 1.341 \times 2 \times 10^5 \times 10^5 = 2.682 \times 10^{10}$$

You may sometimes need an extra step, if the result of the multiplication gets too big. Remember scientific notation writes a number between one and ten times a power of ten. If your multiplication gives you a number greater than ten, you'll have to adjust.

Problem 7:

Multiply $4 \times 10^3 \times 7 \times 10^8$.

Solution:

$$4 \times 10^3 \times 7 \times 10^8 = 4 \times 7 \times 10^3 \times 10^8 = 28 \times 10^{11} = 2.8 \times 10 \times 10^{11}$$

$$= 2.8 \times 10^{12}$$

Notice that 28 was rewritten as 2.8×10 and the tens were multiplied by adding the exponents.

5   DIVIDING LARGE NUMBERS

Division is the reverse of multiplication. Dividing by 1,000 may be as simple as cancelling three trailing zeros. Twenty million divided by one thousand just needs to have zeros crossed out for you to see it equals twenty thousand.

$$\frac{20,000,000}{1,000} = \frac{20,000,\cancel{000}}{1,\cancel{000}} = \frac{20,000}{1} = 20,000$$

Problem 1:

Divide 275 billion by one thousand.

Solution:

$$\frac{275,000,000,\cancel{000}}{1,\cancel{000}} = 275,000,000 \text{ or } 275 \text{ million}$$

Cancel the trailing zeros in the divisor and the same number of zeros in the numerator, if possible. If you don't have the same number of zeros in both, cancel as many as you can.

Problem 2:

How much is two million divided by ten thousand?

Solution:

$$2,000,\cancel{000} \div 10,\cancel{000} = 200$$

Problem 3:

How much is forty billion divided by one hundred thousand?

Solution:

$$40,000,\cancel{000},000 \div 100,\cancel{000} = 400,000$$

Here again, scientific notation can make things easier. Just divide the numbers that are not powers of ten, then divide the powers by keeping the ten and subtracting the exponents.

Problem 4:

Divide 8.4×10^8 by 2.1×10^5.

Solution:

$$\frac{8.4 \times 10^8}{2.1 \times 10^5} = \frac{8.4}{2.1} \times 10^{8-5} = 4 \times 10^3 = 4,000$$

Problem 5:

How much is 850 billion divided by 17 million?

Solution:

$$\frac{850,000,000,000}{17,000,000} = \frac{8.5 \times 10^{11}}{1.7 \times 10^7} = 5 \times 10^4 = 50,000$$

SELF-TEST 17.4

1. Write 543,000,000 in scientific notation

2. Write 8.51×10^6 in standard form.

3. How much is $1,000^4$?

4. How much is $10,000 \times 7,564$?

5. How much is $100,000 \times 56,412$?

6. Multiply 1.2×10^4 by 6×10^3.

7. How much is $2,340 \times 200,000$?

8. Divide 340 million by 1,000.

9. Divide 15 billion by 10,000.

10. How much is 6.5 trillion divided by five hundred thousand?

6   SCIENTIFIC NOTATION FOR VERY SMALL NUMBERS

Scientific notation can also be used to write extremely small numbers in a compact form. About one hundred million atoms would fit in one centimeter, which means each atom is about $\frac{1}{100,000,000}$ or 0.00000001 cm long. Typing those numbers required a lot of concentration. In scientific notation, the approximate length of an atom is 1×10^{-8}, which is a lot easier to type.

Writing a very small number in scientific notation is very similar to the process for large numbers. Let's use the number 0.00000000000007285 as our example.

- Count the number of digits from the first non-zero digit—in this case the seven—back to the decimal point. In this number, that's fourteen places.

0.00000000000007285 But do you notice we're counting in the opposite
14 places

direction than we did with big numbers? We'll signal that by making our
exponent –14.

- Place a decimal point after the first non–zero digit, in this case, 7.
- Multiply 7.285 by 10^{-14}. The exponent of –14 came from the count of digits.
  You can drop all the zeros before the 7.

$$0.00000000000007285 = 7.285 \times 10^{-14}$$

Problem 1:

Write the number 0.00000467 in scientific notation.

Solution:

$$0.00000467 = 4.67 \times 10^{-6}$$
6 places

Problem 2:

Express 0.00000000000000000039 in scientific notation.

Solution:

$$0.00000000000000000039 = 3.9 \times 10^{-19}$$
19 places

To change a small number in scientific notation to standard form, notice
the exponent. If you're changing 6.5×10^{-11} to standard form, you'll need to
move the decimal point 11 places to the left, and right now there's only one
digit to the left of the decimal point. You'll need to write it ten zeros so that
you can move the decimal point eleven steps left.

$$6.5 \times 10^{-11} = 00000000006.5 \times 10^{-11} = .000000000065$$
11 places left

Problem 3:

Express in standard form: 6×10^{-5}

Solution:

$$6 \times 10^{-5} = 0.\,00006$$
5 places left

Problem 4:

Express in standard form: 3.27×10^{-9}

Solution:

$$3.27 \times 10^{-9} = 0.\underbrace{00000000327}_{9 \text{ places left}}$$

The good news is the multiplying and dividing in scientific notation follow the same rules whether you're working with very large or very small numbers.

Problem 5:

Multiply 2×10^{-12} by 4×10^{-11}

Solution:

$2 \times 10^{-12} \times 4 \times 10^{-11} = 8 \times 10^{-23}$

Problem 6:

Multiply 5.2×10^{-8} by 3.7×10^{-12}

Solution:

$5.2 \times 10^{-8} \times 3.7 \times 10^{-12} =$

$5.2 \times 3.7 \times 10^{-8} \times 10^{-12} = 19.24 \times 10^{-20} = 1.924 \times 10 \times 10^{-20} = 1.924 \times 10^{-19}$

Did you remember that scientific notation must begin with a number between 1 and 10? If the negative numbers are giving you trouble, review Chapter 6.

Problem 7:

Divide 5.2×10^{-12} by 2.6×10^{-8}

Solution:

$$5.2 \times 10^{-12} \div 2.6 \times 10^{-8} = 2 \times 10^{-12-(-8)} = 2 \times 10^{-4}$$

Problem 8:

Divide 1.2×10^{-7} by 4.0×10^{-9}

Solution:

$$1.2 \times 10^{-7} \div 4.0 \times 10^{-9} = 0.3 \times 10^{-7-(-9)} = 0.3 \times 10^2 = 3 \times 10^{-1} \times 10^2$$

$$= 3 \times 10 = 30$$

SELF-TEST 17.5

1. Write in scientific notation.

   (a) 5,200,000   (b) 0.000081

2. Write in standard form.

   (a) 5.3×10^5 (b) 7.9×10^{-6}

3. Multiply each pair of numbers.

   (a) 8.4×10^{17} and 2.3×10^5 (b) 1.8×10^{-12} and 3.0×10^{-4}

4. Multiply $2 \times 10^{15} \times 4 \times 10^{-8} \times 3 \times 10^{-9}$

5. Divide the first number by the second.

   (a) 6.3×10^{15} by 2.1×10^8 (b) 1.4×10^{-8} by 7.0×10^{-4}

ANSWERS TO SELF-TEST 17.1

1. 475,500   2. 2,000,000   3. 774,250,000

4. 93,002   5. 346,561,288   6. seventy-five thousand

7. four hundred million, one hundred sixty thousand

8. one hundred forty-five thousand five

9. seven hundred eighty-five million, six hundred thirty-one thousand, seventy-two

10. one hundred million, one hundred seventy-five thousand, two hundred

ANSWERS TO SELF-TEXT 17.2

1. $275,600,000 = 2.756 \times 10^8$

2. $9,240,000 = 9.24 \times 10^6$

3. $63,400,000 = 6.34 \times 10^7$

4. $58,000 = 5.8 \times 10^4$

5. $30,042,500 = 3.00425 \times 10^7$

6. $6.27 \times 10^5 = 6.27000 \times 10^5 = 627,000$

7. $5.84 \times 10^8 = 5.84000000 \times 10^8 = 584,000,000$

8. $7.1 \times 10^6 = 7.100000 \times 10^6 = 7.100,000$

9. $9 \times 10^5 = 9.00000 \times 10^5 = 900,000$

10. $3.01 \times 10^{11} = 3.01000000000 \times 10^{11} = 301.000.000,000$

ANSWERS TO SELF-TEST 17.3

1. 50,000,000,000

2. 400,230,000,000,000

3. 72,954,308,000,000

4. 10,000,000,000,000

5. 502,464,917,000,000

6. one hundred and seventy-five billion

7. eighty billion

8. six hundred seventy-five trillion, five hundred and five billion, seven hundred sixty- eight million, four hundred thousand

9. three trillion

10. two hundred thirty-four trillion, four hundred fifty-nine billion, eight hundred seventy- five million, three hundred sixty-one thousand, eight hundred and twenty-seven

ANSWERS TO SELF-TEST 17.4

1. $543,000,000 = 5.43 \times 10^8$

2. $8.51 \times 10^6 = 8.510000 \times 10^6 = 8,510,000$

3. $1,000^4 = 10^3 \times 10^3 \times 10^3 \times 10^3 = 10^{12} = 1,000,000,000,000$

4. 75,640,000

5. 5,641,200,000

6. $1.2 \times 10^4 \times 6 \times 10^3 = 7.2 \times 10^7 = 72,000,000$

7. $2,340 \times 200,000 = 2.34 \times 10^3 \times 2 \times 10^5 = 4.68 \times 10^8 = 468,000,000$

8. 340,000

9. 1,500,000

10. 6.5 trillion ÷ 500 thousand $= (6.5 \times 10^{12}) \div (5 \times 10^5) = 1.3 \times 10^7 = 13,000,000$

ANSWERS TO SELF-TEST 17.5

1. (a) 5.2×10^6
   (b) 8.1×10^{-5}

2. (a) 530,000
   (b) 0.0000079

3. (a) 1.863×10^{23}
   (b) 5.4×10^{-16}

4. $2.4 \times 10^{-1} = 0.24$

5. (a) 3×10^7
   (b) 2×10^{-5}

18 Algebra Problems

In this chapter we'll set out to find the ages of various people, the identities of groups of numbers, and the composition of various mixtures. What do these different problems have in common? Each involves the search for x, the unknown.

These exercises are some of the traditional applications of basic algebra and will develop your algebraic reasoning. They are a controlled training ground, allowing you to sharpen your skills, and gain the confidence to apply algebra to real questions. So, yes, it is silly to think that someone would know that Tom is twice as old as Mary was three years ago, and not know how old Tom is, but you will practice representing situations in ways that algebra can clarify.

The work of this chapter will rely on Chapter 15, "Solving Simple Equations," Chapter 5, "Mental Math" and Chapter 6, "Positive and Negative Numbers." Be sure you're comfortable with those. Let's get started.

1 AGE PROBLEMS

Let's go back to x, the great unknown. Surely there is some x you've always been curious about—perhaps it's somebody's age. Suppose someone said, "I'm 10 years older than George." If you knew that the person telling you this was 50, it wouldn't be hard to figure out that George is 40. Or, if this person, who you know is 50, said, "I'm twice as old as George," you'd figure out that George is 25.

Believe it or not, you were using algebra to figure out George's age in both instances. If x were George's age and someone who was 50 said he was 10 years older than George, you would have a ready-made equation:

$$x + 10 = 50$$
$$x = 40$$

Let's take this one step further.

Problem 1:

If George's age were still x, the unknown, and someone who is 90 told you he was three times as old as George, write an equation to find George's age.

Solution:

$$3x = 90$$

$$x = 30$$

Those were *easy* ones. We'll try something a little harder. Remember to let x be the unknown.

Problem 2:

If you are 28, and you are 5 years less than three times Alex's age, how old is Alex?

Solution:

$$3x - 5 = 28$$

$$3x = 33$$

$$x = 11$$

If the circumstances of that problem were true, Alex would be 11 years old. It's always wise to go back and check that your answer makes sense in the problem. No one is −43 years old. There can't be $6\frac{2}{3}$ people on your football team. Check your answer.

Here's another one.

Problem 3:

A mother is twice as old as her daughter. Their combined ages equal 60. How old is the mother and how old is the daughter? (Hint: The key here is to figure out what to have x represent. We're going to do it with x representing the daughter's age. You could make x the mother's age, but your equation will have a fraction, which is fine, if you're comfortable with fractions.)

Solution:

Let daughter's age $= x$
Let mother's age $= 2x$

$$x + 2x = 60$$

$$3x = 60$$

$$x = 20$$

$$2x = 40$$

So the mom is 40 and her daughter is 20. That could happen, so let's move on.

Here's another one.

Problem 4:

Alice is twice as old as Jesse. Jesse is twice Marie's age. Their combined ages equal 105. How old is each?

(Hint: Here again, you could set up differently, depending on what you say x represents. Try ignoring the 105 for a moment and just making up ages for Alice, Jesse, and Marie. Who's oldest? Who's youngest? Do you want to have fractions or avoid them? Write down what x stands for, and any other symbols you use.)

Solution:

Let x = Marie's age
Let $2x$ = Jesse's age
Let $2 \times 2x = 4x$ = Alice's age

$$x + 2x + 4x = 105$$
$$7x = 105$$
$$x = 15$$
$$2x = 30$$
$$4x = 60$$

Marie is 15 years old, Jesse is 30, and Alice is 60. You may be tempted to use more than one letter in a problem like this, and there are ways to do that, but they involve working with more than one equation. Try to use just x, if you can.

How are you doing? If you need help in setting up equations, go to Frame 1 of Chapter 15, reread the entire chapter, and then begin this chapter again.

This next one is really pushing it. If you get it right, then pat yourself on the back for a job well done. If you've gotten everything in this section right up to now and you have trouble with this one, don't worry about it. It's a challenge.

Problem 5:

Three years ago, a mother was five times her daughter's age. Three years from now, she will be only three times as old as her daughter. How old are the daughter and the mother today?

(Hint: When a problem gives you information that doesn't immediately fit together, it can help to organize it in a chart. Let's set one up with a column for the daughter and a column for the mother. Make the rows 3 years ago, now, and 3 years from now.)

Solution:

Let x = daughter's age today
Let $x - 3$ = daughter's age 3 years ago
Let $5(x - 3)$ = mother's age 3 years ago
Let $3(x + 3)$ = mother's age 3 years from now

	Daughter	Mother
3 years ago	$x - 3$	$5(x - 3)$
Now	x	
3 years from now	$x + 3$	$3(x + 3)$

Notice that we still have an empty space, the mother's age now. And you may still be feeling lost for an equation. But we do know that right now the mother is three years older than she was three years ago, so we could write her current age as $5(x - 3) + 3$, and that means that three years from now, she'll be $5(x - 3) + 3 + 3$ or $5(x - 3) + 6$, and we know that, three years from now, she'll be three times her daughter's age.

	Daughter	Mother
3 years ago	$x - 3$	$5(x - 3)$
Now	x	$5(x - 3) + 3$
3 years from now	$x + 3$	$5(x - 3) + 6$ or $3(x + 3)$

Therefore:

$$5(x - 3) + 6 = 3(x + 3)$$

$$5x - 15 + 6 = 3x + 9$$

$$5x - 9 = 3x + 9$$

$$5x = 3x + 18$$

$$2x = 18$$

$$x = 9$$

Right now, the daughter is 9 years old, three years ago she was 6 and in three years from now, she'll be 12. Three years ago, the mom was $5 \times 6 = 30$ years old, now she is 33, and three years from now, she'll be 36.

Of course, if this were real life, you could have just asked them their ages, but hopefully the puzzle intrigued you at least a little bit.

SELF-TEST 18.1

1. If Marcia is 4 years short of being twice Linda's age, and the sum of their ages is 107, how old are these women?

2. Ken is three times Janice's age and twice Bill's age. If the combined ages of the three equal 55, how old are Ken, Janice, and Bill?

3. Alan is three times Sheila's age. In 5 years, he'll be just twice her age. How old are Alan and Sheila?

4. Two years ago, Fred was four times Steve's age. Three years from now he will be just three times Steve's age. How old are Fred and Steve today?

FINDING THE NUMBERS

After the workout you had in the last section, this one will be child's play. What we'll be doing here is finding the mystery numbers. When we talk about *consecutive* numbers, we're referring to numbers one after another as you count, for example, 37, 38, and 39. Consecutive numbers differ by 1.

Problem 1:
Three consecutive numbers add up to 24. Find the numbers.

Solution:
Let x be the first number
Let $x + 1$ be the second number
Let $x + 2$ be the third number

$$x + x + 1 + x + 2 = 24$$
$$3x + 3 = 24$$
$$3x = 21$$

$$x = 7$$
$$x + 1 = 8$$
$$x + 2 = 9$$

Problem 2:

Four consecutive numbers add up to 82. Find the numbers.

Solution:

Let x be the first number
Let $x + 1$ be the second number
Let $x + 2$ be the third number
Let $x + 3$ be the fourth number

$$x + x + 1 + x + 2 + x + 3 = 82$$
$$4x + 6 = 82$$
$$4x = 76$$
$$x = 19$$
$$x + 1 = 20$$
$$x + 2 = 21$$
$$x + 3 = 22$$

Pretty easy, huh? Okay, let's get creative.

Problem 3:

Find three consecutive numbers that add to -39.

Solution:

Let $x =$ the first number
Let $x + 1 =$ the second number
Let $x + 2 =$ the third number

$$x + x + 1 + x + 2 = -39$$
$$3x + 3 = -39$$
$$3x = -42$$

$$x = -14$$

$$x + 1 = -13$$

$$x + 2 = -12$$

Did you say -14, -15, and -16? That's a common mistake. If we had gotten $x = 14$, you would have said 14, 15, 16 without a second thought, but if $x = -14$, $x + 1 = -14 + 1 = -13$. Negative numbers can be confusing. Trust the way you set the problem up and check your answers.

Problem 4:

Find four consecutive numbers that sum to -2.

Solution:

Let $x =$ the first number
Let $x + 1 =$ the second number
Let $x + 2 =$ the third number
Let $x + 3 =$ the fourth number

$$x + x + 1 + x + 2 + x + 3 = -2$$

$$4x + 6 = -2$$

$$4x = -8$$

$$x = -2$$

$$x + 1 = -1$$

$$x + 2 = 0$$

$$x + 3 = +1$$

Problem 5:

The sum of three numbers is -16. The second number is four times the first. The third number is 2 more than the second.

(Hint: notice the word consecutive does NOT appear in this problem.)

Solution:

Let $x =$ the first number
Let $4x =$ the second number
Let $4x + 2 =$ the third number

$$x + 4x + 4x + 2 = -16$$

$$9x + 2 = -16$$

$$9x = -18$$
$$x = -2$$
$$4x = -8$$
$$4x + 2 = -6$$

SELF-TEST 18.2

1. Find three consecutive numbers that add up to 42.

2. Find three consecutive numbers that add up to −3.

3. Three numbers add up to 82. The second number is four times as large as the first, and the third is 8 less than the second. Find the numbers.

4. The sum of three numbers is 5. The second number is three times the first. The third number is 10 less than the first. Find the numbers.

5. The sum of four numbers is 60. If the second number is twice as large as the first, the third is 2 less than the second, and the fourth is 3 larger than twice the third, find the numbers.

3   MIXTURE PROBLEMS

You know those cans of mixed nuts that are always about 70 percent peanuts? Well, suppose that we make up our own mixtures and go very easy on the peanuts.

Problem 1:

Let's make up a mixture of 10 pounds of nuts. We'll use some peanuts, which sell for $2 per pound, and some almonds, which sell for $4 per pound. Now if we sell this mixture, we want to charge at least enough to cover the cost of the ingredients, but not so much that it's too expensive for the market. Sometimes we pick our price and adjust the ingredients to make it work. If this nut mixture will sell for $3.50 per pound, how many pounds of peanuts and how many pounds of almonds should we use?

   You could fiddle around with different numbers for a bit, if it helps you get a sense for what's reasonable. Remember we want a total of 10 pounds.

It will probably be faster if you use your algebra. To get started, let $x =$ pounds of something.

Solution:

Let $x =$ pounds of peanuts
Let $10 - x =$ pounds of almonds

You could let x stand for the pounds of almonds and $10 - x$ for the pounds of peanuts. It will still work out, but it's important to write down which you chose, so you remember at the end.

Next step:

What is the value of the ingredients that go into the mixture? Find the value by multiplying the number of pounds times the price per pound. A chart can be helpful here.

	Pounds	Price per Pound	Value
Peanuts	x	$2	$2x$
Almonds	$10 - x$	$4	$4(10 - x)$
Mixture	10	$3.50	$35.00

We want the value of the peanuts and the value of the almonds, combined, to be equal to the value of the mixture. The value of the mixture is what the ten pounds of mixed nuts will sell for, in this case, $35. So, we can set up this equation:

$$\$2x + \$4(10 - x) = \$35$$

$$\$2x + \$40 - \$4x = \$35$$

$$-\$2x + \$40 = \$35$$

$$-\$2x = -\$5$$

$$-2x = -5$$

$$2x = 5$$

$$x = 2\frac{1}{2}$$

$$10 - x = 7\frac{1}{2}$$

The dollar signs remind us of what kind of numbers we're handling, but have no effect on the algebra, so if you want to drop them earlier than we did here, that's fine.

If you're puzzled by our going from $-\$2x = -\$5$ to $\$2x = \$5$, here's what we did: we multiplied both sides by -1. Remember, you can do virtually anything to one side of an equation if you do it to the other side as well. It just got rid of the negative signs, which make some people nervous. You can leave $-\$2x = -\$5$ and just divide both sides by -2, if you prefer.

Now that all the work was done on this last one, you do the work on the next problem.

Problem 2:

We're going to upgrade our mixture. Let's mix cashews ($5 per pound) with pecans ($8 per pound). We'll do a 50-pound mixture that will sell for $6.50 per pound.

Solution:

Let x = pounds of cashews
Let $50 - x$ = pounds of pecans
The value of the cashews is $\$5x$. The value of the pecans is $\$8(50 - x)$, or $\$400 - \$8x$. And the value of the mixture is $50 \times \$6.50$, or $\$325$.

$$\$5x + \$400 - \$8x = \$325$$

$$-\$3x + \$400 = \$325$$

$$-\$3x = -\$75$$

$$\$3x = \$75$$

$$3x = 75$$

$$x = 25$$

$$50 - x = 25$$

Notice that the price of the mixture ($6.50) split the difference between the cashews ($5) and the pecans ($8). You could have saved yourself the trouble of all that work by just saying, since the price was split down the middle, that the mixture must be half cashews and half pecans. As you get used to working with numbers, this type of observation will become second nature.

These problems are so long that we'll go directly to the self-test. However, if you're not sure how to do them, please go back to Frame 3 and go over the problems again.

SELF-TEST 18.3

1. A 20-pound mixture of peanuts and Brazil nuts sells for $6 a pound. If peanuts sell for $3 a pound and Brazil nuts for $7 a pound, how many pounds of peanuts and how many pounds of Brazil nuts are used in the mixture?

2. An 80-pound mixture of shelled walnuts and cashews is sold for $4 a pound. If the walnuts sell for $3 a pound and the cashews for $6 a pound, how many pounds of walnuts and how many pounds of cashews are in the mixture?

3. A 60-pound mixture of peanuts and shelled pistachio nuts sells for $5 a pound. If the peanuts sell for $2 a pound and the pistachios are $8 a pound, how many pounds of pistachios and how many pounds of peanuts are in the mixture?

ANSWERS TO SELF-TEST 18.1

1. Let x = Linda's age
   Let $2x - 4$ = Marcia's age

$$x + 2x - 4 = 107$$
$$3x - 4 = 107$$
$$3x = 111$$
$$x = 37$$
$$2x - 4 = 70$$

Linda is 37 and Marcia is 70.

2. Let x = Janice's age
   Let $3x$ = Ken's age
   Let $\frac{3}{2}x$ = Bill's age

$$x + 3x + \frac{3}{2}x = 55$$
$$4x + \frac{3x}{2} = 55$$
$$\frac{8x}{2} + \frac{3x}{2} = 55$$
$$\frac{11x}{2} = 55$$

$$11x = 110$$

$$x = 10$$

$$3x = 30$$

$$\frac{3}{2}x = 15$$

Janice is 10, Ken is 30, and Bill is 15.

3. Let x = Sheila's age
   Let $3x$ = Alan's age

$$3x + 5 = 2(x + 5)$$

$$3x + 5 = 2x + 10$$

$$3x = 2x + 5$$

$$x = 5$$

$$3x = 15$$

	Steve	Fred
2 years ago	$x - 2$	$4(x - 2)$
Now	x	$4(x - 2) + 2$
In 3 years	$x + 3$	$4(x - 2) + 5$ or $3(x + 3)$

4. Let x = Steve's age now
   Let $4(x - 2)$ = Fred's age 2 years ago
   Let $4(x - 2) + 5$ = Fred's age 3 years from now
   Let $3(x + 3)$ = Fred's age 3 years from now also

$$4(x - 2) + 5 = 3(x + 3)$$

$$4x - 8 + 5 = 3x + 9$$

$$4x - 3 = 3x + 9$$

$$4x = 3x + 12$$

$$x = 12$$

Steve is 12 and Fred is 42.

1. Let x = the first number
   Let $x + 1$ = the second number
   Let $x + 2$ = the third number

$$x + x + 1 + x + 2 = 42$$
$$3x + 3 = 42$$
$$3x = 39$$
$$x = 13$$
$$x + 1 = 14$$
$$x + 2 = 15$$

2. Let x = the first number
   Let $x + 1$ = the second number
   Let $x + 2$ = the third number

$$x + x + 1 + x + 2 = -3$$
$$3x + 3 = -3$$
$$3x = -6$$
$$x = -2$$
$$x + 1 = -1$$
$$x + 2 = 0$$

3. Let x = the first number
   Let $4x$ = the second number
   Let $4x - 8$ = the third number

$$x + 4x + 4x - 8 = 82$$
$$9x - 8 = 82$$
$$9x = 90$$
$$x = 10$$
$$4x = 40$$
$$4x - 8 = 32$$

4. Let x = the first number
   Let $3x$ = the second number
   Let $x - 10$ = the third number

$$x + 3x + x - 10 = 5$$
$$5x - 10 = 5$$
$$5x = 15$$
$$x = 3$$
$$3x = 9$$
$$x - 10 = -7$$

5. Let x = the first number
   Let $2x$ = the second number
   Let $2x - 2$ = the third number
   Let $2(2x - 2) + 3$ = the fourth number

$$x + 2x + 2x - 2 + 2(2x - 2) + 3 = 60$$
$$x + 2x + 2x - 2 + 4x - 4 + 3 = 60$$
$$9x - 3 = 60$$
$$9x = 63$$
$$x = 7$$
$$2x = 14$$
$$2x - 2 = 12$$
$$2(2x - 2) + 3 = 27$$

The first number is 7, the second is 14, the third is 12, and the fourth is 27.

ANSWERS TO SELF-TEST 18.3

1. Let x = pounds of peanuts
   Let $20 - x$ = pounds of Brazil nuts

The value of peanuts in the mixture is $3x. The value of Brazil nuts in the mixture is $7(20 − x), or $140 − $7x. The entire mixture sells for $6 × 20, or $120.

$$\$3x + \$140 - \$7x = \$120$$
$$-\$4x + \$140 = \$120$$
$$-\$4x = -\$20$$
$$\$4x = \$20$$
$$4x = 20$$
$$x = 5$$
$$20 - x = 15$$

2. Let x = pounds of walnuts
   Let $80 - x$ = pounds of cashews
   The value of walnuts in the mixture is $3x. The value of cashews in the mixture is $6(80 − x), or $480 − $6x. The entire mixture sells for $4 × 80, or $320.

$$\$3x + \$480 - \$6x = \$320$$
$$-\$3x + \$480 = \$320$$
$$-\$3x = -\$160$$
$$\$3x = \$160$$
$$3x = 160$$
$$x = 53\frac{1}{3}$$
$$80 - x = 26\frac{2}{3}$$

3. The easy way is to split the difference: 30 pounds of peanuts and 30 pounds of pistachio nuts. We can do this because the price of the mixture is $5, which is halfway between $2 and $8.
   Here's the entire solution for those who worked it out.
   Let x = pounds of peanuts
   Let $60 - x$ = pounds of pistachios

The value of the peanuts is $2x. The value of the pistachios is $8(60 − x), or $480 − $8x. The entire mixture sells for $5 × 60, or $300.

$$\$2x + \$480 - \$8x = \$300$$

$$-\$6x + \$480 = \$300$$

$$-\$6x = -\$180$$

$$\$6x = \$180$$

$$6x = 180$$

$$x = 30$$

$$60 - x = 30$$

19 Interest Rates

When you borrow money, you pay back the money you borrowed plus a bit more, called interest. How much interest you pay is calculated as a percentage of what you borrowed. It's important to understand the different ways interest can be calculated or the interest rate you pay may be a lot higher than you bargained for. When you receive interest for the money you lend out, you may not necessarily receive the rate of interest you think you are. Interest is not a simple concept, but as we go from the simple interest rate to the compound, you'll see the true rate of interest.

1 SIMPLE INTEREST

Simple interest is the term we give to a way of calculating that extra cost of borrowing by one multiplication: the amount you borrow, times the rate of interest, times the amount of time the loan lasts. The amount borrowed is called the principal. The rate is usually given as a percent per year. You'll need to change the percent to a decimal or a fraction to do the calculation. The time is generally in years, because the rate is usually "per year" but if the rate were "per month" you'd measure the time in months. The formula for simple interest is Interest = Principal × Rate × Time or $I = P \times r \times t$.

If you lent $100 for 1 year to a friend, and she paid you 5% interest per year, how much money would she give you when she repaid the loan?

$$I = P \times r \times t = \$100 \times 0.05 \times 1 \text{ year} = \$5.$$

She would repay $100 plus $5 interest, or $105

How much money would she have paid you if the loan had been for 2 years? She would have paid you $110: the $100 principal, plus $10 interest ($5 for each year).

What if she had borrowed $100 for 10 years? Then, using the same 5% rate of simple interest, she would have paid you $150, or $100 principal plus $50 interest ($5 for each of the 10 years that the loan was outstanding).

Now we'll do simple interest loans for parts of a year.

Problem 1:

How much interest is paid on a $200 loan that is made for 6 months at an annual simple interest rate of 6%?

Solution:

The interest rate is 6% per year so you'll need to express 6 months as ½ a year. $I = P \times r \times t = \$200 \times 0.06 \times ½$ year $= \$6$.

The interest paid would be $6, but don't confuse the interest with the amount to be repaid. The principal must be repaid plus interest, so the borrower would repay $206.

Problem 2:

How much interest is paid on a $1,000 loan that is made for 3 months at an annual simple interest rate of 8%?

Solution:

Express 3 months as $\frac{3}{12} = \frac{1}{4}$ of a year and 8% as 0.08. $I = P \times r \times t = \$1,000 \times 0.08 \times \frac{1}{4}$ year $= \$20$.

Eight percent of $1,000 is $80. Since the loan was made for just one-quarter of a year, $20 interest is paid. The total repayment would be $1,020.

SELF-TEST 19.1

Calculate the simple interest paid for each of these problems:

1. $5,000 for 6 months at an annual rate of 8%.

2. $3,000 for 3 years at an annual rate of 7%.

3. $4,000 for 1 year at an annual rate of 9%.

4. $2,000 for one quarter at an annual rate of 12%.

2   COMPOUND INTEREST

Most of us encounter the concept of interest when a financial institution pays us interest for the use of our money via savings or investment. The truth is you're unlikely to see simple interest in that situation.

The more common practice would be to use compound interest, a method that can be described as paying interest on your interest. For example, let's say you have $100 in a savings account with a bank that pays 4% interest compounded quarterly. That was once a common practice, and it meant that at the end of each quarter, the bank would calculate 4% of your average balance for that quarter and add it to your account. To keep it simple, let's say your balance was $100 for every day of the quarter. The bank would pay you $100 × $0.04 \times \frac{1}{4}$ = $1 in interest for that quarter by adding that to your $100. Now you have $101 in your account. If you don't withdraw any money during the next quarter, you'll receive $101 × $0.04 \times \frac{1}{4}$ = $1.01 in interest next quarter. It doesn't sound like much but it will add up over time.

When interest was first compounded quarterly, if you closed your account before the end of a quarter, you got no interest for that quarter, even if you only missed one day. As computers made calculations easier and faster, the banks were able to move to compounding monthly, then daily, and eventually to compounding continuously. (More on continuous compounding later in this chapter.)

Welcome to the magic of compound interest. The more frequently your interest is compounded, the more you earn interest on your interest. It may seem as though the few pennies difference in any one calculation wouldn't matter but over time, more frequent compounding does help.

Compound interest is, as you'll see, a calculation where pencil and paper becomes tedious and you'll feel grateful and totally justified in using a calculator or computer. For some of the problems below, we'll show the pencil and paper version, but for others we'll set up the calculation and then use a calculator.

To make your work easier, we'll give you a formula for compound interest, but it needs a bit of explanation first. There's a logic to the formula, not just magic. For simple interest, you used the formula $I = P \times r \times t$. We're going to start with that, because compound interest is just that calculation, over and over again. Let's think about an amount of money, P, earning a rate of interest per year that we'll call r, but compounded quarterly. At the end of the first quarter, the bank calculates your interest as $I = P \times r \times \frac{1}{4}$. Without more numbers, we can't actually calculate that, but to make it a little tidier, let's write it as $I = P \times \frac{r}{4}$. That's your interest, so the bank credits that to your account, and now you have P + I dollars in your account. This is your new principal after one quarter, so let's call it P_1. But $P_1 = P + I = P + \left(P \times \frac{r}{4}\right)$, which means you have one whole P and a fraction of P, whatever amount P is. Let's write it this way: $P_1 = P + \left(P \times \frac{r}{4}\right) = P \times \left(1 + \frac{r}{4}\right)$. (You can multiply that out using the distributive property to make sure it's equal.)

At the end of the second quarter, interest is calculated again. Your account balance is P_1 and that gets multiplied by r and $\frac{1}{4}$. $I = P_1 \times r \times \frac{1}{4} = P_1 \times \frac{r}{4}$. That gets added on to what you already had and your new principal after two quarters is $P_2 = P_1 + \left(P_1 \times \frac{r}{4} \right) = P_1 \left(1 + \frac{r}{4} \right)$.

Look at what's happening.

Beginning: P

After one quarter: $P_1 = P \left(1 + \frac{r}{4} \right)$

After two quarters: $P_2 = P_1 \left(1 + \frac{r}{4} \right)$

Now a little magic happens. Because we know $P_1 = P \left(1 + \frac{r}{4} \right)$, we can rewrite $P_2 = P_1 \left(1 + \frac{r}{4} \right)$ as $P_2 = P \left(1 + \frac{r}{4} \right) \left(1 + \frac{r}{4} \right)$. If $P_1 = P \left(1 + \frac{r}{4} \right)$ and $P_2 = P \left(1 + \frac{r}{4} \right) \left(1 + \frac{r}{4} \right)$, can you guess what your principal will be after three quarters? It will be $P_3 = P \left(1 + \frac{r}{4} \right) \left(1 + \frac{r}{4} \right) \left(1 + \frac{r}{4} \right)$. If we use exponents to save space, $P_2 = P \left(1 + \frac{r}{4} \right)^2$ and $P_3 = P \left(1 + \frac{r}{4} \right)^3$, the pattern continues for the fourth quarter and beyond.

If P dollars are invested at $r\%$ annual rate compounded t times a year for y years, the amount A at the end will be $A = P \left(1 + \frac{r}{t} \right)^{ty}$.

If interest is compounded quarterly, for 2 years, the $t = 4$, and the $y = 2$, and the exponent becomes 8 because interest will be calculated 8 times, 4 times in each year.

Suppose the annual rate of interest were 12% and interest were paid for 2 years compounded monthly. What is r? What is t? What is y? And what is the exponent? The r would be 0.12, and the t would be 12 (interest is compounded 12 times a year), the y would be 2, and the exponent would be $12 \times 2 = 24$.

The formula won't do us much good unless we apply it to some problems. We'll start with an easy one.

Problem 1:

You put $1,000 in the bank. The bank pays an annual rate of 4% interest, compounded quarterly. You leave your money there for 2 years. How much money do you have in your account at the end of 2 years?

Solution:

Start with the formula. Identify all the different numbers and put them in their places.

$$A = P \left(1 + \frac{r}{t} \right)^{ty}$$

P = $1,000, r = 0.04, t = 4, and y = 2.

$$A = 1{,}000\left(1 + \frac{0.04}{4}\right)^{4\times2}$$
$$= 1{,}000(1 + 0.01)^8$$
$$= 1{,}000(1.01)^8$$

Let's go over just that. The initial principal is $1,000. The annual rate of interest is 4%, which comes to 1% per quarter for 2 years, or 8 quarters.

The rest is up to you and hopefully your calculator. Finding the value of 1.01 to the eighth power with paper and pencil isn't hard, but it would take quite a while. This is a totally appropriate use of a calculator. Your calculator may have a key that lets you type in 1.01, hit the key and type in 8, but if it doesn't you can still type out 1.01 × 1.01 × 1.01 × 1.01 × 1.01 × 1.01 × 1.01 × 1.01. Just don't lose track of how many times you typed it. Go ahead and find A.

$$A = \$1{,}000(1 + .01)^8$$
$$= \$1{,}000(1.01)^8$$
$$\approx \$1{,}000(1.0829)$$
$$\approx \$1{,}082.90$$

We carried out two arithmetic operations here: multiplication with decimals and quick multiplication. If you happened to have forgotten how to do either of these two operations. you'll find instructions on multiplication with decimals in Frame 2 of Chapter 8, "Decimals," and you'll find information on quick multiplication in Frame 1 of Chapter 5, "Mental Math."

Problem 2:

Five hundred dollars is lent at an annual rate of 6% interest for 1 year. The interest is compounded monthly. How much money does the lender end up with (principal plus interest)?

Solution:

$$A = P\left(1 + \frac{r}{t}\right)^{ty}$$
$$= \$500\left(1 + \frac{.06}{12}\right)^{12\times1}$$
$$= \$500(1.005)^{12}$$
$$\approx \$500(1.0617)$$
$$\approx \$530.84$$

$(1.005)^{12} = 1.06167781186$

Round to 1.0617.

Here's another one.

Problem 3:

Ten thousand dollars is lent at an annual rate of 4% for 2 years. The interest is compounded quarterly. How much money does the lender end up with at the end of the year?

Solution:

$$A = P\left(1 + \frac{r}{t}\right)^{ty}$$

$$= \$10,000\left(1 + \frac{.04}{4}\right)^{4\times2}$$

$$= \$10,000(1.01)^8$$

$$= \$10,000 \times 1.082857$$

$$= \$10,828.57$$

$1.01^8 = 1.08285670563$

Round to 1.082857

3   A WORD ON CONTINUOUS COMPOUNDING

Earlier we mentioned that banks moved to something called *continuous compounding*. Of course, it's not actually possible to compound continuously because you'd never be able to do anything else. What's meant by that expression is that mathematicians, always curious, asked what happens as you compound more and more often, and discovered a way to predict the result of compounding so often it is practically continuous. They looked at what happens to the formula $A = P\left(1 + \frac{r}{t}\right)^{ty}$ if you keep the values of P, r, and y the same but make the value of t bigger and bigger. They discovered that they could give a formula for the result when t became really large.

The formula involves a number which, like the number we call π, has a decimal part that doesn't end or repeat, so like π, it gets a name to make it easier to write. This number is called e and it's approximately $2.7182818284590452\ldots$ Your calculator may have a key especially for e. If not, 2.71828 is good enough. We're not going to ask you to do any problems about continuous compounding, but we will give you the formula and show you a comparison of different compounding methods. There's a lot of rounding, so our answers will be approximate but you'll see that continuous compounding earns a little more interest.

If an amount of money P is invested at r% annual interest for y years, compounded continuously, the amount A after y years is

$$A = Pe^{ry}$$

Compare investing $1000 for 10 years at 5% annual interest, compounded quarterly with investing the same amount at the same rate for the same time, compounded continuously.

$$A = P\left(1 + \frac{r}{t}\right)^{ty}$$
$$= 1{,}000\left(1 + \frac{.05}{4}\right)^{4\times10}$$
$$= 1{,}000(1.0125)^{40}$$
$$\approx 1{,}000(1.643619)$$
$$\approx 1{,}643.62$$

$$A = Pe^{ry}$$
$$\approx 1{,}000(2.71828)^{.05\times10}$$
$$\approx 1{,}000(2.71828)^{0.5}$$
$$\approx 1{,}000 \times 1.6487207$$
$$\approx 1{,}648.72$$

The continuous compounding gets you an extra $5.10.

Before you begin Self-Test 19.2, ask yourself one question: "Do I really understand what's going on?" If the answer's no, then go right back to Frame 2.

SELF-TEST 19.2

1. One thousand dollars is put in a bank that pays interest at an annual rate of 8%, compounded quarterly. How much money would be in the account after 3 years?

2. Ten thousand dollars is lent out at 4% annual rate. If interest is compounded quarterly, how much money would the borrower owe the lender after 5 years?

3. One hundred thousand dollars is deposited in the bank. If the bank pays an annual rate of interest of 12%, compounded monthly, how much money is in the account after 1 year?

4 DOUBLING TIME: THE RULE OF 70

How long would it take a nation's population to double if it is increasing at the rate of 1% a year? One hundred years? Not necessarily. How long would it take your money to double if it earns 1% annual interest compounded quarterly? You know how to calculate the amount in an account, and in the continuous compounding example in Frame 3 that 5% annual interest for 10 years wasn't enough to double the $1,000 you started with. What's your guess for how long it would take to turn $1,000 into $2,000? Try out your guess by using the formula. We'll wait. Use 5% interest or 1%, as you wish. We'll be here when you're done.

Ah! You're back. Did you guess right? Too high? Too low? It turns out that there's a clever little rule that can help, called the rule of 70. Now, realize it's not perfect. How long it takes to double your money is going to depend on the interest rate, and how frequently it's compounded, but the rule of 70 will get you close.

The rule of 70 says that the doubling time is equal to 70 divided by your interest rate (just drop the % but don't change to a decimal). If you're earning 5% interest, you'll double your money in about $70 \div 5 = 14$ years. If you invest at 10% annual interest, doubling time will be $70 \div 10 = 7$ years. If the rate is only 1%, it will take about 70 years to double your money.

Remember the question about the population growing at 1% per year? We said it wouldn't necessarily take 100 years to double the population because we knew about the rule of 70. Now lots of things can influence population growth and it's hard, if not impossible, to know if that rate of growth will stay the same, but let's look at the calculation for a population of 10,000 people growing at 1% per year, compounded annually for 70 years.

$$A = P\left(1 + \frac{r}{t}\right)^{ty}$$

$$= 10{,}000\left(1 + \frac{.01}{1}\right)^{1 \times 70}$$

$$= 10{,}000(1.01)^{70}$$

$$\approx 10{,}000(2.0067)$$

$$\approx 20{,}067$$

Doubled, with a few people to spare. You can check for yourself if you want, but we assure you that 69 years is not quite enough.

Problem 1:

If you put $1,000 in the bank at 2% interest, compounded annually, how long would it take your money to double?

Solution:

It would take about 35 years. That's right. All we need to do to find the doubling time is to divide the annual compound rate of increase into 70. So we divide 2 into 70, and we get 35.

Problem 2:

If your body weight were increasing at an annual compound rate of 7% per year, how long would it take for your weight to double?

Solution:

$$70 \div 7 = 10$$

So, it would take 10 years.

The rule of 70 can come in quite handy. And it can, used judiciously, make you look like a whiz with numbers.

SELF-TEST 19.3

How much is the doubling time for each of the following problems?

1. A compound annual rate of 7%

2. A compound annual rate of 5%

3. A compound annual rate of 4%

4. A compound annual rate of 14%

5 DISCOUNTING

When making a loan, the lender may use a practice called *discounting*. Discounting the loan means the lender calculates the interest that must be paid and deducts it from the amount you asked to borrow. Essentially, they're getting paid first. You, on the other hand, find yourself with less than you asked to borrow.

Problem 1:

You take out a $1,000 loan at 7% interest. The bank gives you $930, which is $1,000 less the $70 interest (7% of $1,000 is $70). You agreed to pay 7% interest on $1,000, which is $70, but now the bank has that $70 and you only have $930, not $1,000. Does it feel like you're paying more? How much more? Figure it out.

Solution:

$$\text{interest rate} = \frac{\text{amount of interest}}{\text{loan amount} - \text{amount of interest}}$$

$$\frac{\$70}{\$930} = \frac{7}{93}$$

$$= 7.5\%$$

$$93 \overline{) 7.0000} \quad .0752$$

$$\text{xx}$$
$$-6.51$$
$$\overline{\quad 490}$$
$$-465$$
$$\overline{\quad 250}$$
$$-186$$

Now this may not seem like such a big discrepancy, but remember that a difference of half a percent can really mean a lot more interest on a large loan.

Problem 2:

How much is the actual interest rate on a $4,000 loan that a bank discounts at 12%?

Solution:

12% of $4,000 is $480. You get $4,000 - $480 = $3520.

$$\frac{\$480}{\$3,520} = \frac{48}{352}$$

$$= \frac{12}{88}$$

$$= \frac{3}{22}$$

$$= 13.6\%$$

$$22 \overline{) 3.000} \quad .136$$

$$\text{xx}$$
$$-2\ 2$$
$$\overline{\quad 80}$$
$$-66$$
$$\overline{\quad 140}$$
$$-132$$
$$\overline{\quad 8}$$

SELF-TEST 19.4

What is the actual interest rate on each of the following bank discounts? Each loan is for 1 year.

1. A $5,000 loan discounted at 8%

2. A $2,000 loan discounted at 10%

3. A $10,000 loan discounted at 7%

ANSWERS TO SELF-TEST 19.1

1. $\$5,000 \times 0.08 \times \frac{1}{2} = \$200$

2. $\$3,000 \times 0.07 \times 3 = \$630$

3. $\$4,000 \times 0.09 \times 1 = \$360$

4. $\$2,000 \times 0.12 \times \frac{1}{4} = \$60$

ANSWERS TO SELF-TEST 19.2

1. $A = P\left(1 + \frac{r}{t}\right)^{ty}$

   $= \$1,000\left(1 + \frac{.08}{4}\right)^{4 \times 3}$

   $= \$1,000(1.02)^{12}$

   $= \$1,000 \times 1.26824$

   $= \$1,268.24$

   $1.02^{12} = 1.26824179456$
   Round to 1.26824.

2. $A = P\left(1 + \frac{r}{t}\right)^{ty}$

   $= \$10,000\left(1 + \frac{.04}{4}\right)^{4 \times 5}$

   $= \$10,000(1.01)^{20}$

   $= \$10,000 \times 1.220190$

   $= \$12,201.90$

   $1.01^{20} = 1.22019003995$
   Round to 1.220190.

3. $A = P\left(1 + \frac{r}{t}\right)^{ty}$

   $= \$100,000\left(1 + \frac{.12}{12}\right)^{12 \times 1}$

   $= \$100,000(1.01)^{12}$

   $= \$100,000 \times 1.1268250$

   $= \$112,682.50$

   $1.01^{12} = 1.12682503013$
   Round to 1.1268250

ANSWERS TO SELF-TEST 19.3

1. 10 years   2. 14 years   3. 17.5 years   4. 5 years

ANSWERS TO SELF-TEST 19.4

1. $\dfrac{\$400}{\$4,600}$ $= \dfrac{4}{46}$

     $= \dfrac{2}{23}$

     $= 8.7\%$

2. $\dfrac{\$200}{\$1,800}$ $= \dfrac{2}{18}$

     $= \dfrac{1}{9}$

     $= 11.1\%$

3. $\dfrac{\$700}{\$9,300}$ $= \dfrac{7}{93}$

     $= 7.5\%$

20 Rate, Time, and Distance

Most people enjoy travel. They may not be fond of commuting to work or sitting stuck in rush hour traffic, but Americans love to pack up the car and hit the open road or catch a bus, train, or plane. Whether it's a trip to visit family or friends, a vacation getaway, or just to see the world, travel raises three basic questions: (1) how far? (2) how fast? and (3) how long? Here we'll cover how to answer these questions.

1 THE MAGIC FORMULA

One of the greatest formulas of all time is distance = rate × time, or, among the in-crowd, $d = r \times t$. We'll use this formula and its two variations, $r = d \div t$ and $t = d \div r$, to solve every problem in this chapter.

Don't rush to memorize all three of these. They're rearrangements of the same formula, and the rearranging is accomplished by a little bit of basic algebra: multiply or divide both sides of an equation by the same number. If $d = r \times t$, then dividing both sides by t and cancelling $\frac{d}{t} = \frac{r \times t}{t}$ gives you $\frac{d}{t} = r$ or $r = d \div t$. Start with $d = r \times t$ and divide both sides by r, and you get $t = d \div r$.

In Frame 3, we use $d = r \times t$ to tell us how far we've gone. In Frame 4, we learn to calculate how fast we've gone by using $r = d \div t$. Frame 5 helps us to determine how long the trip took from $t = d \div r$.

And finally, we do some mixing and matching—that is, we'll mix all three types of problems in one section and you'll match each one with the proper formula and solve it. You can start with the formula variant that fits the problem and just do the arithmetic, or start with the basic version and deal with the algebra when it comes up.

2 THE TERMS

We need to go over each of the terms, d (distance), t (time), and r (rate). Distance may be expressed in terms of inches, feet, yards, meters, miles, or kilometers. Don't worry, we'll be using miles virtually all of the time.

Rate, or the rate of speed, is usually stated in miles per hour. In the United States, we're so casual that we often don't bother to state "miles per hour," or even m.p.h. On our highways, the signs usually say, "Speed Limit: 60."

Time may be stated in seconds, minutes, hours, days, and so forth. Since we'll be dealing here with travel problems, the time element will generally be expressed in terms of hours.

The important part is to be sure that your measurements are in compatible units. Multiplying 55 miles per hour times 30 seconds won't tell you much. Dividing 65 miles by 15 minutes won't produce miles per hour. Miles per minute, perhaps, but is that what you want? Read the units, not just the numbers.

3   FINDING DISTANCE

We'll begin with an obvious problem. How far does a plane traveling at a speed of 500 m.p.h. go in 1 hour? It goes 500 miles. What distance does it go in 2 hours? Obviously, it goes 1,000 miles.

This gives us our formula: distance = rate × time. In this last instance, distance = 500 m.p.h. × 2 hours.

Problem 1:

A car averaging 55 m.p.h. reaches its destination in $3\frac{1}{2}$ hours. How far did it travel?

Solution:

$$d = r \times t$$
$$= 55 \times 3.5$$
$$= 192.5 \text{ miles}$$

Since the problems we are doing involve rates expressed in miles per hour, we don't bother to write "m.p.h."; since time is expressed in hours, we don't bother to write "hours."

Problem 2:

Jessica leaves her house at 9 A.M. and walks steadily at a rate of $3\frac{1}{2}$ m.p.h. until noon. How far did she walk?

Solution:

$$d = r \times t$$
$$= 3\frac{1}{2} \times 3 = \frac{7}{2} \times \frac{3}{1} = \frac{21}{2}$$
$$= 10\frac{1}{2} \text{ miles}$$

Problem 3:

Joshua leaves on a trip at 10 A.M. He stops 1 hour for lunch and arrives at his destination at 4:30 P.M. If he averaged 50 m.p.h., how far did he drive?

From 10 A.M. to 4:30 P.M. is $6\frac{1}{2}$ hours, but he stopped for one hour to have lunch, so he was only driving for $5\frac{1}{2}$ hours.

Solution:

$$d = r \times t$$

$$= 50 \times 5\frac{1}{2} = \frac{\overset{25}{\cancel{50}}}{1} \times \frac{11}{\underset{1}{\cancel{2}}}$$

$$= 275 \ \text{miles}$$

Problem 4:

Elizabeth started driving at 3 P.M., averaging 55 m.p.h. She stopped for 1 hour to have supper at 6 P.M. After supper she drove at 50 m.p.h. until 9:30 P.M. How far did she travel?

Solution:

We'll break down this problem into two parts: (1) the trip before supper, from 3 P.M. to 6 P.M. and (2) the trip after supper from 7 P.M. to 9:30 P.M.

$$(1) \ \ d = r \times t$$

$$= 55 \times 3$$

$$= 165 \ \text{miles}$$

$$(2) \ \ d = r \times t$$

$$= 50 \times 2.5$$

$$= 125 \ \text{miles}$$

$$165 + 125 = 290 \text{miles}$$

SELF-TEST 20.1

1. If a boat traveled for $6\frac{1}{2}$ hours at 25 m.p.h., how far would it travel?

2. If you left on a trip at 9:30 A.M. and drove steadily at 52 m.p.h. until 1:45 P.M., how far did you drive?

3. If a plane took off at 3 P.M. and flew at 450 m.p.h. until 4:30 P.M., and then took off again at 5:30 P.M. and flew at 500 m.p.h. until 7:30 P.M., how far did it fly?

4. John left work at 5 P.M. and walked at 3 m.p.h. until 6:30 P.M., then he took a bus the rest of the way home. How far does John live from work if the bus traveled at 15 m.p.h. and left him off at his door at 7:00 P.M.?

4   FINDING RATE

If we know how far someone traveled and we know how long the trip took, we can find that person's average rate of speed. All we need to do is plug the distance and time data into the formula, $r = d \div t$.

Remember this is just a variant of the distance formula. The formula $r = \frac{d}{t}$ is derived by dividing both sides of $d = r \times t$ by t.

Problem 1:

A group of cyclists left on a 100-mile trip at noon and arrived at their destination at 8 P.M. What was their average rate of speed in miles per hour?

Solution:

$$r = \frac{d}{t}$$

$$r = \frac{100}{8}$$

$$r = 12.5 \ \text{m.p.h.}$$

Problem 2:

Cindy drove from Washington, D.C., to a suburb of Boston—a distance of 470 miles. She left Washington at 8 A.M. and arrived at her destination at 8 P.M. She stopped along the way for a total of $2\frac{1}{2}$ hours for meals and rest. What was her average rate of speed?

Solution:

$$r = \frac{d}{t}$$

$$r = \frac{470}{12 - 2\frac{1}{2}} = \frac{470}{9.5} = \frac{\overset{940}{\cancel{4700}}}{\underset{19}{\cancel{95}}} = 49\frac{9}{19}$$

$$r \approx 49.5 \ \text{m.p.h}$$

Problem 3:

Susan walked uphill from 9 A.M. until noon, covering a distance of 7 miles. At 1 P.M. she began her return trip, downhill, which took until 3 P.M. What was her average rate of speed when walking uphill? What was her average speed walking downhill? What was her average rate of speed for the entire trip?

Solution:

$$\text{Uphill: } r = \frac{d}{t} \qquad\qquad \text{Downhill: } r = \frac{d}{t}$$

$$r = \frac{7}{3} \qquad\qquad\qquad\qquad r = \frac{7}{2}$$

$$r = 2\frac{1}{3} \text{ m.p.h.} \qquad\qquad r = 3.5 \text{ m.p.h.}$$

$$\text{Overall: } r = \frac{d}{t}$$

$$r = \frac{14}{5}$$

$$r = 2.8 \text{ m.p.h.}$$

How about making some use of our algebra?

Problem 4:

A passenger train starts from New York, heading to San Francisco, and a freight train starts from San Francisco, heading toward New York. The distance between these cities is 2,800 miles. The passenger train is moving at twice the rate of the speed of the freight train. After 10 hours they are still 1,600 miles apart. Find the rate of each train.

Solution:

Let r = the rate of the freight train
Let $2r$ = the rate of the passenger train

Since distance = $r \times t$, the distance traveled by the freight train is $r \times 10$, or $10r$. The distance traveled by the passenger train is $2r \times 10$, or $20r$.

The two trains covered a total distance of $10r + 20r$ or $30r$. If there is still 1,600 miles between them, they covered a total of $2,800 - 1,600 = 1,200$ miles.

$$30r = 1,200$$

$$r = 40$$

$$2r = 80$$

Here's another way to tackle this problem. To use the $r = \frac{d}{t}$ formula to solve this problem, let r = the combined rate of the passenger and freight trains. We know that they covered a distance of 1,200 miles in 10 hours. Substitute these numbers into the formula:

$$r = \frac{d}{t}$$

$$r = \frac{1,200}{10}$$

$$r = 120$$

This leaves us with figuring out how fast each train was traveling. So, let x = the rate of the freight train and $2x$ = the rate of the passenger train.

$$x + 2x = 120$$

$$3x = 120$$

$$x = 40$$

$$2x = 80$$

Which method is better? We like the first one better because it's shorter. But either method is okay because they both give us the right answers.

Problem 5:

Two cars started driving toward each other at 8 A.M. on the New York State Thruway. One started from Albany and the other from Buffalo, 300 miles away. One car drove 10 miles an hour faster than the other car. At 10 A.M., the cars were 80 miles apart. How fast were the cars going?

Solution:

Let r = the rate of the slower car
Let $r + 10$ = the rate of the faster car
Distance traveled by slower car in 2 hours = $2r$
Distance traveled by faster car in 2 hours = $2(r + 10) = 2r + 20$
Distance traveled by both cars = $2r + 2r + 20 = 4r + 20$
Distance traveled by both cars in 2 hours = $300 - 80 = 220$ miles

$$220 = 2r + 2r + 20$$

$$220 = 4r + 20$$

$$200 = 4r$$

$$50 = r$$

$$60 = r + 10$$

SELF-TEST 20.2

1. If a plane went 2,700 miles in $4\frac{1}{2}$ hours, what was its average rate of speed?

2. If Bob left his house at 7 A.M. and rode his bicycle 53 miles by 11 A.M., what was his average rate of speed?

3. Doreen ran 8 miles in 45 minutes. What was her average rate of speed in miles per minute? What was her speed in miles per hour?

4. If Joseph drove from home to work, a distance of 30 miles, in half an hour, and returned by a different route in 45 minutes, what was his average rate of speed?

5. Two planes begin to fly toward each other from points 8,000 miles apart, and one plane is flying one and a half times as fast as the other. If after 3 hours they are still 5,000 miles apart, how fast is each plane flying?

6. Two trains start off from the same station and travel in opposite directions. After 4 hours they are 680 miles apart. If the first train is traveling at a rate of 20 m.p.h. faster than the second train, at what speeds are the trains traveling?

5 | FINDING TIME

If we know how far people traveled and we know how fast they went per hour, we can find how long their trips took. All we need to do is plug the distance and speed data into the formula: $t = \frac{d}{r}$. Remember, $t = \frac{d}{r}$ is derived by starting with the distance formula and dividing both sides by r.

Problem 1:

Michelle drove 300 miles at an average rate of speed of 40 m.p.h. How long did her trip take?

Solution:

$$t = \frac{d}{r}$$

$$t = \frac{300}{40} = 7.5$$

$$t = 7.5 \ \text{hours}$$

Problem 2:

Mark drove to work, a distance of 40 miles, at 50 m.p.h. He drove home from work at a speed of 45 m.p.h. How long did it take him to drive to and from work?

Solution:

We need to divide the problem into two parts: (1) the time it took to drive to work and (2) the time it took to drive home.

$$(1)\ t = \frac{d}{r} \qquad\qquad (2)\ t = \frac{d}{r}$$

$$t = \frac{40}{50} = \frac{4}{5}\ \text{hours} \qquad\qquad t = \frac{40}{45} = \frac{8}{9}\ \text{hours}$$

$$(1) + (2) = \frac{4}{5} + \frac{8}{9} = \frac{4 \times 9}{5 \times 9} + \frac{8 \times 5}{9 \times 5} = \frac{36}{45} + \frac{40}{45} = \frac{76}{45} = 1\frac{31}{45}\ \text{hours}$$

Now suppose we want to convert $1\frac{31}{45}$ hours to hours and minutes. Just multiply the fraction of an hour, $\frac{31}{45}$, by 60, the number of minutes in an hour.

$$1\frac{31}{45}\ \text{hours} = 1\ \text{hour},\ \frac{31}{45} \times \frac{60}{1}\ \text{minutes}$$

$$= 1\ \text{hour},\ \frac{31}{\underset{3}{45}} \times \frac{\overset{4}{60}}{1}\text{minutes}$$

$$= 1\ \text{hour},\ \frac{124}{3}\ \text{minutes}$$

$$= 1\ \text{hour},\ 41\frac{1}{3}\ \text{minutes}$$

One-third of a minute is one-third of 60 seconds, or 20 seconds, so $1\frac{31}{45}$ hours = 1 hour, 41 minutes, and 20 seconds.

Problem 3:

Ms. Jones left home at 7 A.M. and rode her bicycle to work. She covered the distance of 8 miles at a speed of 12 m.p.h. She walked home at a speed of 4 m.p.h. How long did her trip to work and her trip home take her?

Solution:

This is a two-part question.

Trip to work:

$$t = \frac{d}{r}$$

$$t = \frac{8}{12}$$

$$= \frac{2}{3} \text{ hour}$$

Trip home:

$$t = \frac{d}{r}$$

$$t = \frac{8}{4}$$

$$= 2 \text{ hours}$$

Total commuting time: $2\frac{2}{3}$ hours or 2 hours 40 minutes

Problem 4:

Mr. and Mrs. Greenblatt left their home at 10 A.M., traveling in opposite directions. If Mrs. Greenblatt traveled at 48 m.p.h. and Mr. Greenblatt traveled at 44 m.p.h., at what time will they be 322 miles apart?

Solution:

Let's combine their two rates of speed:

$$48 + 44 = 92 \text{ m.p.h}$$

$$t = \frac{d}{r}$$

$$= \frac{322}{92} = \frac{161}{46} = 3\frac{23}{46}$$

$$= 3.5 \text{ hours}$$

They will be 322 miles apart 3.5 hours after 10 A.M. or at 1:30 P.M.

SELF-TEST 20.3

1. Nicole walked at an average rate of speed of $3\frac{1}{2}$ m.p.h. and covered a distance of 14 miles. How long did she walk?

2. Harry ran to work, a distance of 10 miles, at 8 m.p.h. He took the bus home. If the bus traveled at a speed of 20 m.p.h., how long did it take Harry to get to and from work?

3. Two trains left the station traveling in opposite directions. One train was traveling at a rate of 70 m.p.h., and the other was traveling at the rate of 80 m.p.h. When they were 825 miles apart, for how long had they been traveling?

4. Alice got on Highway 80 at Omaha and headed west, averaging 55 m.p.h. John got on Highway 80 at the same time heading east from Omaha, traveling at an average rate of 52 m.p.h. When they were 535 miles apart, how long had they been traveling?

6   RATE, TIME, AND DISTANCE PROBLEMS

Nothing new is covered in this section. All we're going to do here is give you a self–test of all three types of problems. The trick is to figure out which type of problem you're doing and then to write down the proper formula and solve it.

If you're not confident about doing the $d = r \times t$ problems, you should reread Frame 3 and retake Self–Test 20.1. If $r = \frac{d}{t}$ problems leave you at all uneasy, then you should reread Frame 4 and retake Self–Test 20.2. And if you're not entirely clear on $t = \frac{d}{r}$ problems, then reread Frame 5 and retake Self–Test 20.3.

SELF-TEST 20.4

1. If a bus traveling at an average speed of 46 m.p.h. reaches its destination in $2\frac{1}{2}$ hours, how far has it gone?

2. If a person needs $3\frac{1}{3}$ hours to walk 10 miles, how fast does she walk in m.p.h.?

3. If a plane flying at 500 m.p.h. flies 1,200 miles, how long does this trip take?

4. Max ran for $1\frac{1}{2}$ hours at a pace of 10 m.p.h. After resting, he walked back at a pace of 4 m.p.h. How long did his entire trip take?

5. Laura drove to work, a distance of 60 miles, covering the distance in 75 minutes. She took the bus home. This took 2 hours. What was her average rate of speed?

6. A plane flies at 550 m.p.h. for 45 minutes and 600 m.p.h. for $1\frac{1}{2}$ hours. How far did the plane fly?

7. Achilles and Hector decide to race each other around the world. They start at noon from Troy. Achilles, heading due east, runs at an average speed of 8 m.p.h. Hector heads due west, running at 7.5 m.p.h. At what time will they be 93 miles apart?

8. Two trains start off from the same station and travel in opposite directions. After 6 hours they are 840 miles apart. If the first train is traveling at a rate of 20 m.p.h. faster than the second train, at what speeds are the trains traveling?

7 | SPEED LIMIT PROBLEMS

Back in 1973, when the price of oil quadrupled and gasoline shortages (or the fear of them) caused motorists to wait in line for hours at gas stations all around the country, the federal government passed a law. A national speed limit of 55 miles per hour was set. Not only would this conserve gasoline, since fuel economy falls sharply at speeds beyond 55 m.p.h., but this measure would also promote highway safety.

Fuel conservation was attained and highway deaths fell dramatically. But after a few years, a lot of people began to complain to Congress that at a speed of 55 m.p.h., it took too long to cover great distances. And furthermore, it was reasoned, there were plenty of stretches of highway in sparsely populated areas where speed limits could be safely raised.

Since the late 1980s, Congress has allowed the states to set their own limits in accordance with local driving conditions. This action gave rise to a whole new set of rate × time = distance problems. Here they come.

Problem 1:

How much time would you save by driving for 100 miles at 60 m.p.h. rather than 55 m.p.h.?

Solution:

$$t = \frac{d}{r}$$

$$\text{At } 55: t = \frac{100}{55} = \frac{20}{11} = 1\frac{9}{11} \text{ hours}$$

$$\text{At } 60: t = \frac{100}{60} = \frac{10}{6} = \frac{5}{3} = 1\frac{2}{3} \text{ hours}$$

Time saved $= 1\frac{9}{11} - 1\frac{2}{3} = 1\frac{27}{33} - 1\frac{22}{33} = \frac{5}{33}$ of an hour or about $\frac{5}{\overset{}{\underset{11}{\cancel{33}}}} \times \frac{\overset{20}{\cancel{60}}}{1} =$

$\frac{100}{11} \approx 9$ minutes

Problem 2:

How much time would you save if you drove 20 miles at 60 m.p.h. instead of at 50 m.p.h.?

Solution:

$$t = \frac{d}{r}$$

$$\text{At } 50 : t = \frac{20}{50}$$

$$= \frac{2}{5} \text{ hour} = 24 \text{ minutes}$$

$$\text{At } 60 : t = \frac{20}{60}$$

$$= \frac{1}{3} \text{ hour} = 20 \text{ minutes}$$

Time saved $= 4$ minutes

Problem 3:

How much farther would you get if you drove for half an hour at 75 m.p.h. rather than 55 m.p.h.?

Solution:

$$d = r \times t$$

$$\text{At } 55: d = 55 \times \frac{1}{2}$$

$$= 27\frac{1}{2} \text{ miles}$$

$$\text{At } 75: d = 75 \times \frac{1}{2}$$

$$= 37\frac{1}{2} \text{ miles}$$

$$37\frac{1}{2} - 27\frac{1}{2} = 10 \text{ miles}$$

One last question. This will appear to be a trick question, but if you put the right numbers into the right formula, you'll get it right.

Problem 4:

How much time would you save if you drove 1 mile at 65 m.p.h. rather than 55 m.p.h.?

Solution:

$$t = \frac{d}{r}$$

at 55: $t = \frac{1}{55}$ hour $\approx$ $55\overline{)1.00000}^{.01818}$ at 65: $t = \frac{1}{65}$ hour $\approx$ $65\overline{)1.00000}^{.01538}$

Time saved = 0.01818 − 0.01538 = 0.00280 hours = 0.168 minutes = 10.08 seconds.

If you didn't carry this out as far as done here, or even if you didn't get it right, don't lose any sleep over it. Did you get at least two out of the first three problems right? Then go to Self-Test 20.5. If you didn't, please go back to Frame 1 of this chapter. Remember, it's always better the second time around.

SELF-TEST 20.5

1. How much time would you save by driving for 250 miles at 60 m.p.h. rather than 55 m.p.h.?

2. How much time would you save if you drove 55 miles at 60 m.p.h. rather than at 55 m.p.h.?

3. How much time would you save if you drove for 10 miles at 75 m.p.h. rather than at 55 m.p.h.?

4. How much farther would you get if you drove for 10 minutes at 75 m.p.h. rather than at 55 m.p.h.?

5. How much farther would you get if you drove for 20 minutes at 60 m.p.h. rather than at 55 m.p.h.?

1. $d = r \times t$

    $= 25 \times 6.5$

    $= 162.5$ miles

2. $d = r \times t$

    $= 52 \times 4.25$

    $= 221$ miles

3. a. $d = r \times t$

    $= 450 \times 1.5$

    $= 675$ miles

   b. $d = r \times t$

                   $= 500 \times 2$

                   $= 1,000$ miles

    $675 + 1,000 = 1,675$ miles

4. a. $d = r \times t$

    $= 3 \times 1.5$

    $= 4.5$ miles

   b. $d = r \times t$

                   $= 15 \times .5$

                   $= 7.5$ miles

    $4.5 + 7.5 = 12$ miles

ANSWERS TO SELF-TEST 20.2

1. $r = \dfrac{d}{t}$

$$= \dfrac{2{,}700}{4.5} = \dfrac{\overset{5{,}400}{\cancel{27{,}000}}}{\underset{9}{\cancel{45}}} = \dfrac{\overset{600}{\cancel{5{,}400}}}{\underset{1}{\cancel{9}}}$$

$$= 600 \text{ m.p.h.}$$

2. $r = \dfrac{d}{t}$

$$= \dfrac{53}{4} = 13\dfrac{1}{4}$$

$$= 13\dfrac{1}{4} \text{ m.p.h.}$$

3. $r = \dfrac{d}{t}$

$$= \dfrac{8}{3/4} = \dfrac{8}{1} \times \dfrac{4}{3}$$

$$= \dfrac{32}{3} = 10\dfrac{2}{3}$$

$$= 10\dfrac{2}{3} \text{ m.p.h.}$$

4. $r = \dfrac{d}{t}$

$$= \dfrac{60}{1\frac{1}{4}} = \dfrac{60}{\frac{5}{4}} = \dfrac{\overset{12}{\cancel{60}}}{1} \times \dfrac{4}{\underset{1}{\cancel{5}}}a$$

$$= 48 \text{ m.p.h.}$$

5. Let r = speed of slower plane
   Let $1\frac{1}{2}r = \frac{3}{2}r$ = speed of faster plane

$$d = r \times t$$

$$d = r \times 3 = 3r = \text{distance traveled by slower plane in 3 hours}$$

$$d = \frac{3}{2}r \times 3 = \frac{9}{2}r = \text{distance traveled by faster plane in 3 hours}$$

Distance traveled by both planes in 3 hours $= 3r + \frac{9}{2}r = \frac{6}{2}r + \frac{9}{2}r = \frac{15}{2}r$

Distance traveled by both planes in 3 hours = 8,000 − 5,000 = 3,000

$$3,000 = \frac{15}{2}r$$

$$6,000 = 15r$$

$$400 = r$$

$$600 = \frac{3}{2}r$$

6. Let r = speed of slower train
   Let $r + 20$ = speed of faster train

$$d = r \times t$$

Distance traveled by slower train in 4 hours $d = r \times 4 = 4r$
Distance traveled by faster train in 4 hours $d = (r + 20)4 = 4r + 80$
Distance traveled by both trains in 4 hours = $4r + 4r + 80 = 8r + 80$
Distance traveled by both trains in 4 hours = 680

$$680 = 8r + 80$$

$$600 = 8r$$

$$75 = r$$

$$95 = r + 20$$

ANSWERS TO SELF-TEST 20.3

1. $t = \dfrac{d}{r}$

$$t = \frac{14}{3.5} = \frac{\overset{28}{\cancel{140}}}{\underset{7}{\cancel{35}}} = 4 \text{ hours}$$

2. (1) $t = \dfrac{d}{r}$ 　　　　　　　　(2) $t = \dfrac{d}{r}$

$\quad = \dfrac{10}{8}$ 　　　　　　　　　　$\quad = \dfrac{10}{20}$

$\quad = 1\dfrac{1}{4}$ hours 　　　　　　　$\quad = \dfrac{1}{2}$ hour

$$(1) + (2) = 1\frac{1}{4} + \frac{1}{2} = \frac{5}{4} + \frac{2}{4} = \frac{7}{4} = 1\frac{3}{4} \text{ hours}$$

3. $t = \dfrac{d}{r}$

   r = combined rate for 1 hour

   $\quad = 80 + 70 = 150$

   $t = \dfrac{825}{150} = \dfrac{33}{6} = \dfrac{11}{2} = 5\dfrac{1}{2}$ hours

4. $t = \dfrac{d}{r}$

   r = combined rate for 1 hour

   $\quad = 55 + 52$

   $\quad = 107$

   $t = \dfrac{535}{107}$

   $\quad = 5$ hours

ANSWERS TO SELF-TEST 20.4

1. $d = r \times t$

   $d = 46 \times \dfrac{5}{2}$

   $\quad = 23 \times 5$

   $\quad = 115$ miles

2. $r = \dfrac{d}{t}$

   $r = \dfrac{10}{3\frac{1}{3}}$

   $\quad = \dfrac{10}{1} \div \dfrac{10}{3}$

   $\quad = \dfrac{10}{1} \times \dfrac{3}{10}$

   $\quad = 3$ m.p.h.

3. $t = \dfrac{d}{r}$

$t = \dfrac{1,200}{500}$

$= \dfrac{12}{5}$

$= 2.4$ hours or 2 hours 24 minutes

4. First part of trip: $= 1\dfrac{1}{2}$ hours

Second part of trip: $t = \dfrac{d}{r} = \dfrac{15}{4} = 3\dfrac{3}{4}$ hours

$1\dfrac{1}{2} + 3\dfrac{3}{4} = \dfrac{3}{2} + \dfrac{15}{4} = \dfrac{6}{4} + \dfrac{15}{4} = \dfrac{21}{4} = 5\dfrac{1}{4}$ hours

5. $r = \dfrac{d}{t}$

$r = \dfrac{120}{3\dfrac{1}{4}}$

$= \dfrac{120}{1} \div \dfrac{13}{4}$

$= \dfrac{120}{1} \times \dfrac{4}{13}$

$= \dfrac{480}{13}$

$r \approx 36.9$ m.p.h.

$$
\begin{array}{r}
36.9 \\
13\overline{)480.0} \\
\text{x x} \\
-39 \\
\hline
90 \\
-78 \\
\hline
12\ 0 \\
-11\ 7 \\
\hline
3
\end{array}
$$

6. a. $d = r \times t$

$d = \dfrac{550}{1} \times \dfrac{3}{4}$

$= \dfrac{1,6550}{4}$

$= 412.5$ miles

b. $d = \dfrac{600}{1} \times \dfrac{3}{2}$

$= \dfrac{1,800}{2}$

$= 900$ miles

$a + b = 412.5 + 900 = 1,312.5$ miles

7. $t = \dfrac{d}{r}$

Combined rate for Achilles and Hector: 15.5 m.p.h.

$$t = \frac{93}{15.5}$$

$$15.5\overline{)93} \;=\; 155\overline{)930} \;=\; 31\overline{)186} \begin{array}{c} 6 \\ \end{array}$$
$$\underline{-186}$$

$$t = 6 \text{ hours}$$

8. Let r = speed of slower train
   Let $r + 20$ = speed of faster train
   Distance covered by slower train in 6 hours = $r \times t = r \times 6$, or $6r$
   Distance covered by faster train in 6 hours = $r \times t = (r + 20)6 = 6r + 120$
   Combined distance covered in 6 hours by both trains = $6r + 6r + 120$
   = $12r + 120$
   Given: Distance covered by both trains in 6 hours = 840 miles

$$840 = 12r + 120$$

$$720 = 12r$$

$$60 = r$$

$$80 = r + 20$$

ANSWERS TO SELF-TEST 20.5

1. $t = \dfrac{d}{r}$

   at 55: $t = \dfrac{250}{55} = \dfrac{50}{11} = 4\dfrac{6}{11} \approx 4.55$

   at 60: $t = \dfrac{250}{60} = \dfrac{25}{6} = 4\dfrac{1}{6} \approx 4.17$

   $t \approx 4.55 - 4.17$

   ≈ 0.38 hour

   $t \approx 22.8$ minutes

2. $t = \dfrac{d}{r}$

   at 55: $t = \dfrac{55}{55} = 1$ hour

   at 60: $t\dfrac{55}{60} = \dfrac{11}{12}$

   $t = 1 - \dfrac{11}{12} = \dfrac{1}{12}$ hour = 5 minutes

3. $t = \dfrac{d}{r}$

at 55: $t = \dfrac{10}{55} = \dfrac{2}{11} \approx 0.18$

at 75: $t = \dfrac{10}{75} = \dfrac{2}{15} \approx 0.13$

$0.18 - 0.13 = 0.05$ hour $= 0.05 \times 60$ minutes $= 3$ minutes

4. $d = r \times t$

at 55: $d = 55 \times \dfrac{1}{6} = \dfrac{55}{6} = 9\dfrac{1}{6}$ miles

$12\dfrac{1}{2} - 9\dfrac{1}{6} =$

$12\dfrac{3}{6} - 9\dfrac{1}{6} = 3\dfrac{2}{6} = 3\dfrac{1}{3}$

$d = 3\dfrac{1}{3}$ miles

at 75: $d = 75 \times \dfrac{1}{6} = \dfrac{75}{6} = 12\dfrac{1}{2}$ miles

5. $d = r \times t$

at 55: $d = 55 \times \dfrac{1}{3} = 18\dfrac{1}{3}$ miles

$d = 20 - 18\dfrac{1}{3} = 1\dfrac{2}{3}$ miles

at 60: $d = 60 \times \dfrac{1}{3} = 20$ miles

21 Personal Finance

As individual consumers, we are called upon to use arithmetic and simple algebra almost every day. Sometimes we do mathematical calculations without even being conscious that we are doing them. And other times, we are all too aware that it is mathematics that we are trying to do.

In this chapter, we have taken up six types of numerical problems that most of us encounter in our day-to-day lives. They all have one thing in common—percentages.

MARK-DOWN PROBLEMS

Everybody looks for bargains. Advertisements such as "Everything must go," "Prices slashed by 50% and more," and "Lost our lease" attract bargain hunters like honey attracts bees. Let's see how much you're actually saving.

Problem 1:

A dress is marked down from $99 to $59. By what percent has the price been cut?

Solution:

$$\text{percentage change} = \frac{\text{change}}{\text{original price}} = \frac{40}{99}$$

$$99 \overline{)40.000} = 40.4\%$$

.404

−396

400

−396

Problem 2:

You will receive a $500 rebate on a $9,000 car. What percentage of the original price do you get back?

Solution:

$$\text{percentage change} = \frac{\text{rebate}}{\text{original price}}$$

$$= \frac{\$500}{\$9,000}$$

$$9000\overline{)500} = 90\overline{)5}$$

$$= 18\overline{)1.0000} \!\!\!\!\begin{array}{l} .0555 \\ \end{array} = 5.6\%$$

$$\begin{array}{r} -90 \\ \hline 100 \\ -90 \\ \hline 100 \\ -90 \end{array}$$

Problem 3:

A pair of shoes originally priced at $39.95 was marked down by 20%. What is the new price?

Solution:

$$\begin{array}{r} \$39.95 \\ \times .20 \\ \hline \$7.9900 = \$7.99 \end{array} \qquad \begin{array}{r} \$39.95 \\ -7.99 \\ \hline \$31.96 \end{array}$$

How are you doing? If you have gotten at least two out of these three right, then go to the next problems. If not, then you should review Frame 2 of Chapter 14 and return to the beginning of this chapter.

And now for something a little different. And a little harder.

Problem 4:

A jacket is marked down by 25% to $50. What was its original price?

Solution:

Let x = the original price

$$x - .25x = \$50$$

$$.75x = \$50$$

$$x = \frac{\$50}{.75}$$

$$.75\overline{)\$50} = 75\overline{)\$5,000}$$

$$= 3\overline{)20^20.^20.^20^20} \!\!\!\!\begin{array}{l} 6\,6.\,6\,6\,6 \\ \end{array}$$

Problem 5:

A bicycle was marked down by 40% to $120. What was its original price?

Solution:

Let x = the original price

$$x - .4x = \$120$$

$$.6x = \$120$$

$$x = \frac{\$120}{.6} = \frac{\$1200}{6} = \$200$$

If you got both problems right, then go to Self-Test 21.1. And if you got them wrong, then you need to review some of your algebra. Remember x, and all the things you may let it represent? Return to Frame 1 of Chapter 18 and reread the entire chapter. Then go back to Problem 4 in this section.

SELF-TEST 21.1

1. A sofa is marked down from $599 to $299. By what percentage has the price been cut?

2. An auto dealer is offering a $1,000 rebate on a $14,000 car. What percentage of the original price do you get back?

3. A dress that was originally $150 was marked down by 40%. What is the new price?

4. A living room set is marked down by 35% to $750. What was its original price?

5. A bedroom set was marked down by 60% to $975. What was its original price?

Which Is the Better Deal?

Who is your long-distance carrier? Maybe I can get you to switch.

Friendly Fones offers you a flat rate of 10 cents a minute, any time, day or night. Chat Line claims to have an even better deal—12 cents a minute, any time, day or night. But your first 30 minutes each month are free. Who is offering the better deal? See if you can work it out.

Solution:

Using Friendly Fones, your first 30 minutes cost you $3.00 (30 × $.10). But your first 30 minutes are free using Chat Line. If you talk for only 30 minutes a month, Chat Line is definitely the way to go.

But after the first 30 minutes, Chat Line costs you 12 cents a minute, while Friendly Fones is just 10 cents a minute. For every additional minute you use Friendly Fones, you're saving 2 cents. How long beyond the initial 30 minutes would you need to talk to make up the entire $3?

The answer is 150 minutes. Okay, then, which long distance carrier gives you the better deal?

If you talk for more than three hours (30 minutes plus 150 minutes), then Friendly Fones is the better deal. But if you talk less than three hours a month, you should definitely go with Chat Line.

2   SALES TAX PROBLEMS

When you pay sales tax, you never have to worry about calculating it because that's the job of the seller. But you're the one who pays the tax. In fact, most people don't even pay attention to how much they're paying for goods or services and how much they're being charged in sales tax. In New York City, for example, there's a sales tax of 8.875% on most items—clothing (over $110), restaurant meals, books, furniture, movie admissions. In New Jersey, just across the river, they charge only 6.625%, and most clothing is tax exempt. So maybe it pays to drive over to New Jersey to shop at retail outlets. Not only do you save on your sales tax, but you can pick up some real bargains. Let's figure out how much you would save.

Problem 1:

If a dress in New York City were priced as $129 and a store in New Jersey had the same dress for $89, how much would you save? Did you say $40? Guess again. You would save even more, because you need to figure out the sales tax that you would have paid in New York. In New Jersey, there's no sales tax on most clothing.

Solution:

$$
\begin{array}{r}
\$129 \\
\times\,.0875 \\
\hline
645 \\
903 \\
1032 \\
\hline
\$11.2875
\end{array}
$$

You would be saving $11.29 in tax, in addition to the $40 price difference. So you would save a total of $51.29. On the other hand, you'd have to pay for gas and tolls, or public transportation.

Problem 2:

If there is a sales tax of 6%, how much tax would you pay on a used car that was priced at $2,500?

Solution:

$$
\begin{array}{r}
\$2,500 \\
\times\ .06 \\
\hline
\$150.00
\end{array}
$$

Now we'll get a bit more involved.

Problem 3:

How much would the original price of an item be if you paid a total (including taxes) of $102.90 and the sales tax rate was 5%?

Solution:

First, you need to find the price of the item that you purchased before the tax was added.

Let x = the original price

$$x + .05x = \$102.90$$

$$1.05x = \$102.90$$

$$x = \frac{\$102.90}{1.05} = \frac{10290}{105} = \frac{3 \times 5 \times 7 \times 98}{3 \times 5 \times 7}$$

$$= \$98$$

If you didn't spot the cancelling, it would be reasonable to use a calculator.

Problem 4:

How much would the original price be if you paid a total (including taxes) of $520 for a sofa and the sales tax rate were 4%?

Solution:

Let x = original price of sofa

$$x + .04x = \$520$$

$$1.04x = \$520$$

$$x = \frac{\$520}{1.04} = \frac{52000}{104} = \frac{5 \times 100 \times 104}{104}$$

$$= \$500$$

If you got these last two right, go to Self-Test 21.2. If you got them wrong, please return to Chapter 18, which is a review of algebra, and read the entire chapter. The whole trick with these problems is to let x represent the unknown. Then everything else will fall into place. When you've finished Chapter 18, return to Frame 2 of this chapter.

SELF-TEST 21.2

1. If the sales tax on a $450 purchase were 4½%, how much tax would you pay?

2. If the sales tax rate in North Dakota were 7% and the sales tax in South Dakota were 3%, how much money would a person save by buying a $9,000 car in South Dakota?

3. How much would the original price be if you paid a total (including taxes) of $350 for a kitchen set and the sales tax rate were 5%?

4. How much would the original price be if you paid a total (including taxes) of $94.50 for a dress and the sales tax were 5%?

3   CREDIT CARDS

The following section is written as a public service, especially for credit card holders who are paying high interest rates. Do you ever read the fine print on the back of your bill? Do you know how much interest you are being charged on your unpaid balance? Well, you're about to find out.

Problem 1:

Suppose you make purchases this month totaling $1,050 dollars and the bank asks you to make a minimum payment of $50. So what do you do? You pay $50. Then the bank charges you, say, 17.5% interest on the unpaid balance. This annual rate is translated into a monthly rate of 1.458%. How much interest do you pay?

Solution:

$$\begin{array}{r} .01458 \\ \times\ \$1000 \\ \hline \$14.58000 = \$14.58 \end{array}$$

You just paid $14.58 for the privilege of carrying a $1,000 balance on your credit card for 1 month.

Problem 2:

You have an outstanding monthly balance of $2,075, and there's a minimum payment due of $75. If the annual interest rate on your balance is 19% and you pay just the minimum, how much interest do you owe after 1 month?

Solution:

$$\text{monthly interest rate} = \frac{\text{annual rate}}{12}$$

$$= \frac{19\%}{12}$$

$$= 1.583\%$$

$$\begin{array}{r} .01583 \\ \times\ \$2000 \\ \hline \$31.6600 = \$31.66 \end{array}$$

Problem 3:

If you owe $47 interest on a monthly credit card balance of $3,000, what is the annual and monthly interest rates that you must pay?

Solution:

$$\frac{\$47}{\$3,000} = 3,000 \overline{)\,47.000}$$

$$= 1.567\% \text{ monthly}$$

$$\begin{array}{r} 1.567 \\ \times\ 12 \\ \hline 3134 \\ 1567 \\ \hline 18.804 = 18.8\% \text{ annually} \end{array}$$

Problem 4:

If you owe $64 interest on a monthly credit card balance of $4,000, what is the annual and monthly interest rates you must pay?

Solution:

$$\frac{\$64}{\$4,000} = \frac{\$16}{\$1,000}$$

$$= 1.6\% \text{ monthly}$$

$$\begin{array}{r} 1.6 \\ \times\ 12 \\ \hline 32 \\ 16 \\ \hline 19.2\% \text{ annually} \end{array}$$

To put all of this into perspective, if you happened to have a bank balance of $1,000 and you had an average credit card balance of $1,000, you would end up paying 16% or 18% interest on your card debt while receiving just 5% interest for your bank deposit. And if you maintained an account at the same bank that issued your credit card, you would have the privilege of borrowing your own money and paying somewhere between 11% and 13% interest for this privilege.

At the website bankrate.com, and others, you can find a calculator that will help you determine how long it will take to pay off your credit card balance and how much interest you will pay in the process. Experimenting with the calculator can help you make a plan to get out of debt.

SELF-TEST 21.3

1. How much interest would you pay on a credit card balance of $3,000 in 1 month if the annual rate of interest were 18%?

2. How much interest would you pay on a credit card balance of $2,500 if the annual rate of interest were 15%?

3. If you owe $50 interest on a monthly credit card balance of $3,300, what are the annual and monthly interest rates that you must pay?

4. If you owe $70 interest on a monthly credit card balance of $4,900, what are the annual and monthly interest rates that you must pay?

4   FEDERAL INCOME TAX

Have you ever heard anyone say, "I can't afford to take on any more work because it would put me in a higher tax bracket"? Have you ever heard anyone say that being in a higher tax bracket means that you actually lose money by working more? What those folks are suggesting is that because earning more will cause them to be taxed at a higher rate, so much of their additional income will go to taxes that it's just not worth it. The only way that statement could be literally true is if they were taxed at a rate over 100%, and there is no income level with that tax rate.

In the 1940s, there was a 90% tax bracket, but never one of 100% or more. The current maximum tax rate is 37%. Today, in 2021, there are seven tax brackets: 10%, 12%, 22%, 24%, 32%, 35% and 37%. Exactly how much you pay depends on several factors, including your income, the number of dependents you have, your deductions, and more.

What we're going to do here is work out a few hypothetical income tax problems. The federal income tax brackets are adjusted every year depending on the inflation rate, so we'll look at the 2021 brackets. A chart may help, and be aware that we rounded these numbers to make things easier.

	Taxable income in the range:						
Married couple	Less than $20,000	$20,000 to $81,000	$81,000 to $173,000	$173,000 to $330,000	$330,000 to $419,000	$419,000 to $628,000	Over $628,000
Tax Bracket	10%	12%	22%	24%	32%	35%	37%

Income isn't the only factor, of course. Everyone is entitled to certain deductions and exemptions. Two such deductions, which we'll take up in the next section, are mortgage interest and property taxes. These exemptions and deductions are subtracted from income *before* you calculate your taxes. But calculating your tax is not necessarily a one-step task. If you're a single taxpayer with $50,000 of taxable income, you don't just take 22% of $50,000. If you do, you'll be giving the IRS more money than it asks for. Your tax will be 10% of the first $10,000 plus 12% of the income over $10,000 but not over $41,000 and 22% on the last $9,000. The IRS instructions provide helpful charts so you don't have to calculate all that, but we will do it in the problems so you can see how it works.

If you're really interested in determining your federal income tax on your own, you can download form 1040 from IRS.gov. There are some instructions on the form, and you can also download the 111-page instruction book. There are also websites that will let you enter your information and will fill out the form and even file it for you. If you want to learn more about Federal income tax, there are books like *Economics: A Self-Teaching Guide*, 2nd edition (Wiley, 1999) by Steve Slavin.

Problem 1:

George Johnson reports his annual salary of $32,000, and exemptions and deductions that come to $20,000. His boss decides to double his salary to $64,000. If Johnson's exemptions and deductions remain the same, how much tax does he pay? How does this compare to what he paid before the salary increase? Has his after-tax income actually doubled?

Solution:

George Johnson's original situation: Total income ($32,000) − deductions and exemptions ($20,000) = taxable income ($12,000).

10% of 10,000 + 12% of ($12,000 − $10,000) = 0.10($10,000) + 0.12($2,000) = $1,000 + $240 = $1,240.

After tax income: $32,000 − $1,240 = $30,760.

His new situation: Total Income ($64,000) − deductions and exemptions ($20,000) = taxable income ($44,000). His $44,000 will be taxed at 10% for the first $10,000, 12% of the next $31,000 and 22% of the last $3,000.

10% of 10,000 + 12% of $31,000 + 22% of $3,000 = 0.10($10,000) + 0.12($31,000) + 0.22($3,000) = $1,000 + $3,720 + $660= $5,380.

After tax income: $64,000 − $5,380 = $58,620. His tax bill increases by $5,380 − $1,240 = $4,140 and his after-tax income increases $58,620 − $30,760 = $27,860. While this is a significant increase, his after-tax income has not doubled. $58,620 ÷ $30,760 = 1.9057.

Problem 2:

The Lees used to earn $85,000 between them, but they both got promotions and now earn $125,000. They are entitled to $33,000 in deductions and exemptions. How much tax did they pay before the promotions? How much did they pay after the promotions?

Solution:

Before promotions: Total income ($85,000) − deductions and exemptions ($33,000) = taxable income ($52,000).

10% of $20,000 + 12% of $32,000 = $2,000 + $3,840 = $5,840 in taxes.

After promotions: Total income ($125,000) − deductions and exemptions ($33,000) = taxable income ($92,000). The first $20,000 will be taxed at 10%, the next $61,000 at 12% (that bracket ends at $81,000), and that will leave $11,000 taxed at 22%.

Ten percent of $20,000 + 12% of $61,000 + 22% of $11,000 = $2,000 + $7,320 + $2,420 = $11,740. Again, were the Lees better off or worse off after their promotions? We can see that their higher earnings pushed them from the 12% bracket into the 22% bracket and that their taxes rose from $5,840 to $11,740.

How much take-home pay did they receive before the promotions? $85,000 − $5,840 = $79,160. How much after? $125,000 − $11,740 = $113,260

So their after-tax income rose from $79,160 to $113,260 even after the IRS took its cut.

This should demonstrate that your after-tax income will never decrease if your before-tax income rises—even if you end up in a higher tax bracket.

SELF-TEST 21.4

For the following problems, use tax rates and tax brackets shown in the taxable income chart in Frame 4.

1. The Guptas had a combined taxable income of $90,000 (before deductions and exemptions of $25,000). How much taxes do they pay?

2. Chris was earning $100,000 before taxes. She took $37,000 in deductions and exemptions and is a single taxpayer. How much tax does she pay?

5 MORTGAGE INTEREST AND TAXES

Mortgages are too complex to be discussed very thoroughly in a book of this nature. We can barely begin the job of confusing you, but any banker will be happy to let you know all about fixed and variable rates, adjustable rates, maximum rate changes, prepayments, points, balloon payments, home equity loans, and second mortgages. Many of them have gone to great expense to print mortgage brochures that are virtually incomprehensible.

All we want to do here is stress two things: (1) the role interest plays in your mortgage payments; and (2) the tax advantages of owning rather than renting. After that, it will be up to you to go into your friendly neighborhood bank and make your best deal.

Let's set up a 3.5%, fixed rate, 30-year, $400,000 mortgage. Suppose you wanted to find out four things:

1. your monthly payments;

2. how much money you will pay the bank over the life of the mortgage;

3. how much of this payment will be interest; and

4. how much of your first monthly payment will be interest and how much will be paid on the principal.

Before we even begin, a confession. You can go to the bankrate.com website, and others, and find calculators that will answer these questions. We can give you a formula to let you calculate the mortgage payment by hand, or probably by hand and calculator, if you want to, but you might want to take advantage of the available automation, at least for the monthly payment.

Here's the formula for the monthly payment M:

$$M = P \times r \times \left[\frac{(1+r)^n}{(1+r)^n - 1} \right]$$

where P = principal loan amount, r = *monthly* interest rate, and n = number of months required to repay the loan.

Here's what the calculation looks like for $P = \$400,000$, $r = \frac{0.035}{12} = 0.00292$ and $n = 12 \times 30 = 360$.

$$M = P \times r \times \left[\frac{(1+r)^n}{(1+r)^n - 1} \right]$$

$$= 400,000 \times 0.00292 \times \left[\frac{1.00292^{360}}{1.00292^{360} - 1} \right]$$

If ever there was a moment to grab your calculator, this would be it.

$$M = 400,000 \times 0.00292 \times \left[\frac{1.00292^{360}}{1.00292^{360} - 1} \right]$$

$$= 400,000 \times 0.00292 \times \left[\frac{2.8532871993}{1.8532871993} \right]$$

$$= 400,000 \times 0.00292 \times 1.5395811175$$

$$= 1798.23$$

This calculation of \$1,798 doesn't include real estate taxes or homeowners or mortgage insurance, all of which many lenders include in your monthly payment, so be prepared for a higher payment when you do your budgeting.

So, however you find it, the monthly payment would be about \$1,798. Using that information, you can find the answers to questions 2 (how much money you will pay the bank over the life of the mortgage) and 3 (how much of this payment will be interest). Go ahead and make your calculations. I'll wait right here.

What did you get? For the amount of money you will pay the bank over the life of the mortgage, all you need to do is multiply \$1,798 by 360, which is the number of months in 30 years. That gives us \$647,280. To determine how much of that will be interest, just subtract the \$400,000 principal and you'll get \$247,280.

The last question, how much of your first monthly payment will be interest and how much will be principal, was a question really best left to the bank's calculators. For the first payment, the interest would be about \$1,168, and the payment on the principal about \$630. Each payment will be a slightly different breakdown. As time goes on, more of the payment goes to principal and less to interest. What happens, of course, is that because you're paying back such a small proportion of your loan each month, you're paying almost entirely interest during the first few years of repayment. But as you continue to pay off the principal, the interest payments become a smaller and smaller proportion of the monthly payment.

Another important question concerns the amount of your average monthly interest payment over the life of the loan. It is found by dividing total interest payments of $247,280 by 360. That gives us $686.88. This number is very important for tax purposes because mortgage interest payments are deductible on your federal income tax, as is property tax, both with certain limitations. These two considerations usually tip the scales in favor of owning rather than renting.

Every deduction you can come up with offsets some taxable income. What owning a home does is give the taxpayer thousands of dollars of income tax deductions. Exactly how many thousands depends on two things: (1) the mortgage interest paid and (2) the local property taxes paid.

Problem 1:

What would be the amount of tax deductions you could claim if you bought a home at the terms previously noted—a $400,000, 30-year mortgage at 3.5%? Assume also a $4,000 deduction for local property taxes.

Solution:

Your deductions would come to $8,242.56 for the mortgage interest ($686.88 × 12) plus $4,000 for the property taxes for a total of $12,242.56.

And how much do those deductions reduce your taxes if you're in the 24% bracket?

$$\$12,242.56 \times 0.24 = \$2,938.21$$

Let's try one more.

Problem 2:

If you were in the 24% bracket and you had monthly mortgage interest payments of $950 and local property taxes of $5,500: (1) how much in tax deductions would you get, and (2) how much would your taxes be reduced?

Solution:

1. $950 × 12 = $11,400 + $5,500 = $16,900

2. $16,900 × .24 = $4,056

SELF-TEST 21.5

1. If you are paying a bank a total of $350,000 on a $150,000 mortgage over a 20-year period, how much are your monthly mortgage interest payments? If you are also paying property taxes of $5,000 and are in the 24% federal income tax bracket, how much of a tax deduction do you get for owning a home and by how much do you reduce your taxes?

2. If you are paying a bank a total of $450,000 on a $200,000 mortgage over a 30-year period, how much are your monthly mortgage interest payments? If you are also paying property taxes of $9,000 and are in the 24% federal income tax bracket, how much of a tax deduction do you get for owning a home and how much do you reduce your taxes?

ANSWERS TO SELF-TEST 21.1

1. $$\text{percentage change} = \frac{\text{change}}{\text{original price}}$$

   $$= \frac{\$300}{599}$$

   $$= 50.1\%$$

   $$\begin{array}{r} .5008 \\ 599\overline{)300.0000} \\ \text{x x x} \\ -299\ 5 \\ \hline 5000 \\ -4792 \\ \hline 208 \end{array}$$

2. $$\text{percentage change} = \frac{\$1,000}{14,000}$$

   $$= \frac{1}{14}$$

   $$= 7.1\%$$

   $$\begin{array}{r} .071 \\ 14\overline{)1.000} \\ \text{x} \\ -98 \\ \hline 20 \\ -14 \\ \hline 6 \end{array}$$

3. $$\begin{array}{r} \$150 \\ \times.4 \\ \hline \$60 \end{array} \qquad \begin{array}{r} \$150 \\ -60 \\ \hline \$90 \end{array}$$

4. Let x = original price

   $$x - .35x = \$750$$

   $$.65x = \$750$$

   $$x = \frac{\$750}{.65} \approx \$1153.846$$

5. Let x = original price

$$x - .6x = \$975$$

$$.4x = \$975$$

$$4x = \$9,750$$

$$x = \frac{\$9,750}{4}$$

$$= \$2,437.50$$

ANSWERS TO SELF-TEST 21.2

1. $450
      $\times .045$
      ‾‾‾‾‾‾‾
      2 250
      18 00
      ‾‾‾‾‾‾‾
      $20.250

2. $9000
      $\times .04$
      ‾‾‾‾‾
      $360

3.   Let x = original price

$$x + .05x = \$350$$

$$1.05x = \$350$$

$$x = \frac{\$350}{1.05}$$

$$= \$333.33$$

4.   Let x = original price

$$x + .05x = \$94.50$$

$$1.05 = \$94.50$$

$$x = \frac{\$94.50}{1.05}$$

$$= \$90$$

ANSWERS TO SELF-TEST 21.3

1. $\dfrac{18\%}{12} = \dfrac{3}{2} = 1.5\%$

$$\begin{array}{r} \$3000 \\ \times\,.015 \\ \hline 15\,000 \\ 30\,00 \\ \hline \$45.000 = \$45.000 \end{array}$$

2. $\dfrac{15\%}{12} = 1\dfrac{3}{12} = 1\dfrac{1}{4}\%$

$$\begin{array}{r} 2500 \\ \times\,.0125 = 1.25\% \\ \hline 1\,2500 \\ 5\,000 \\ 25\,00 \\ \hline 31.2500 = \$31.25 \end{array}$$

3. $\dfrac{\$50}{\$3,300} = \dfrac{5}{330} = \dfrac{1}{66}$

$= 1.515\%$

$$\begin{array}{r} 1.515\% \text{ monthly} \\ \times 12 \\ \hline 2030 \\ 1515 \\ \hline 17.180 = 17.18\% \text{ annually} \end{array}$$

4. $\dfrac{\$70}{\$4,900} = \dfrac{1}{70}$

$= 1.429\%$ monthly

$= 17.148\%$ annually

$$\begin{array}{r} 1.429\% \text{ monthly} \\ \times 12 \\ \hline 2\,858 \\ 14\,29 \\ \hline 17.148\% \text{ annually} \end{array}$$

ANSWERS TO SELF-TEST 21.4

1. $\$90,000 - 25,000 = \$65,000$
    $0.10(\$20,000) + 0.12(\$45,000) = \$2,000 + \$5,400 = \$7,400$

2. $\$100,000 - 37,000 = \$63,000$
    $0.10(\$10,000) + 0.12(\$31,000) + 0.22(\$22,000)$
    $= \$1,000 + \$3,720 + \$4,840$
    $= \$9,560$

ANSWERS TO SELF-TEST 21.5

1. $\dfrac{\text{total paid} - \text{loan amount}}{\text{number of payments}} = \dfrac{\$350,000 - \$150,00}{12(20)}$

$= \dfrac{\$200,000}{240}$

$= \dfrac{\$20,000}{24}$

$= \dfrac{\$10,000}{12}$

$= \dfrac{\$5,000}{6}$

$= \$833.33$ monthly payment

Tax deduction:

$\$833.33 \times 12 = \$10,000 + \$5,000 = \$15,000$

$\$15,000 \times .24 = \$3,600$

2. $\dfrac{\text{total paid} - \text{loan amount}}{\text{number of payments}} = \dfrac{\$450,000 - \$200,00}{12(30)}$

$= \dfrac{\$250,000}{360}$

$= \dfrac{\$25,000}{36}$

$= \dfrac{\$12,500}{18}$

$= \dfrac{\$6,250}{9}$

$= \$694.44$

Tax deduction:

$\$694.44 \times 12 = \$8,333.28$

$\$8,333.28 + \$9,000 = \$17,333.28$

$\$17,333.28 \times .24 = \$4,159.99$

22 Business Math

Businesses, like individual consumers, are enmeshed in percentages. Salespeople usually earn commissions, which are specified percentages of sales. Manufacturers offer retailers discounts, which again are calculated on a percentage basis. And finally, there are the profits that are calculated as a percentage of sales and as a percentage of investment.

In this chapter, we'll use ideas from Chapter 10, "Ratios and Proportions," Chapter 14, "Percentages," and Chapter 15, "Solving Simple Equations." We'll make reference back to mark-downs in Chapter 21, "Personal Finance."

1 COMMISSIONS

Most salespeople work on commission—that is, their earnings depend on their sales. In most cases, they have a base salary but earn additional money as a percentage of the sales they make during the pay period. The more they sell, the more they earn.

Commission arrangements vary widely—there are straight commissions, salaries plus commissions, draws against commissions, graduated commissions, salaries plus quota-bonus commissions, and even more exotic combinations.

We'll confine ourselves mainly to straight commissions and salary plus commission arrangements.

Problem 1:

If a saleswoman is paid a straight 10% commission, how much would she earn on sales of $28,515?

Solution:

$$\$28,515 \times 0.10 = \$2,851.50$$

Problem 2:

How much would the saleswoman earn in 3 months if her monthly sales were $25,500, $16,300, and $30,400 and her 10% commission is in addition to a monthly salary of $3,000?

Solution:

Salary for 3 months: $9,000	Commissions on 3 months of sales;
Sales for 3 months;	$72,000
$25,500	× 0.10
16,300	$7,220
+30,400	
$72,200	Total earned: $9,000 + $7,220 = $16,220.

Problem 3:

How much would a salesman on a 4% commission earn in a year if his quarterly sales were $355,200, $289,710, $216,930, and $401,740?

Solution:

$355,200 $1,263,580
289,710   ×.04
216,930 $50,543.20
+401,740
$1,263,580

Problem 4:

A saleswoman is paid 12% commission on her sales to new customers and 8% on sales to regular customers. If she wrote up $6,000 in sales to new customers and $15,000 in sales to regular customers, how much money did she earn in commissions?

Solution:

$6,000 $15,000
× .12   × .08
$720.00 $1,200.00 $720 + $1,200 = $1,920

Let's move into real estate commissions.

Problem 5:

Suppose a real estate agency charges tenants 15% of their annual rent to find them apartments. If that fee is split evenly between the agency and the salesperson, how much does the salesperson make if she finds someone an apartment that costs $1500 per month?

Solution:

Annual rent: $1,500 × 12 = $18,000

Total fee: 0.15 × $18,000 = $2,700

Salesperson's share of the fee:
$2,700 ÷ 2 = $1,350

Problem 6:

How much would a real estate salesman working for the same company earn by renting someone an apartment that costs $1,750 per month?

Solution:

Annual rent: $1,750 × 12 = $21,000

Total fee: 0.15 × $21,000 = $3,150

Salesperson's share of the fee: $3,150
÷ 2 = $1,575

Problem 7:

Finally, we'll sell a whole house. The agency charges the seller a 4% fee, which it splits evenly with the salesperson. How much does the salesperson earn on the sale of a $650,000 house?

Solution:

Take a shortcut. Half of 4% is 2%,

$$\begin{array}{r} \$650,000 \\ \times\ .02 \\ \hline \$13,000.00 \end{array}$$

SELF-TEST 22.1

Find the commissions earned by each of these salespeople:

1. A monthly salary of $2,500 and a 15% commission on the month's sales of $40,000.

2. An 8% commission on sales of $5,500, $4,300, and $6,800.

3. A 10% commission is earned on sales to new customers, and a 5% commission is paid on sales to regular customers. A salesperson wrote up sales of $67,500 to new customers and $137,500 to regular customers.

4. A real estate agency charges tenants 18% of their annual rent to find them apartments. If that fee is split evenly between the agency and the salesperson, how much does the salesperson make if she finds someone an apartment that costs $2,500 per month?

5. If the fee is split evenly between the real estate agent and the agency, how much would a salesperson earn if a $850,000 house is sold and the agency gets a 5% fee?

2   MARK-UPS

Mark-ups are just the opposite of mark-downs, which we encountered in Chapter 21, "Personal Finances." Some cynics say that every time an item on sale is marked *down*, it had previously been marked *up*. And they're right! Because the mark-up, usually between 20% and 50% of cost, is the way a business firm is able to cover its overhead (operating expenses not directly involved in manufacturing the product) and, hopefully, make a profit.

Problem 1:

Suppose you were running an audio store. If it makes sounds, you sell it. Let's say that your overhead—salaries, rent, utilities, advertising, insurance, and everything else—comes to $10,000 a month. Suppose you sold 1,000 units a month. These units cost you $50 each, or $50,000. What would your mark-up be? Ten percent? Twenty percent? More? Less? At 20% you'd just break-even: $50,000 × 0.20 = $10,000. Twenty percent of the cost of the product would be $10,000, which would just cover your overhead. Unless you're in business for the fun of it, you're going to have to have a higher mark-up. How much of a mark-up would you need to show a $5,000 profit?

Solution:

You want your mark-up to produce $10,000 to cover your overhead and $5,000 more for profit, so you need a percent of $50,000 that equals $15,000.

$15,000 is what percent of $50,000? $\frac{\$15,000}{\$50,000} = \frac{15}{50} = \frac{3}{10} = 30\%$ You'll need a mark-up of 30%.

Problem 2:

How much of a mark-up would you need to show a $20,000 profit?

Solution:

You want a mark-up that produces $10,000 for overhead + $20,000 profit or $30,000. What percent of $50,000 will be $30,000? $\frac{\$30,000}{\$50,000} = \frac{3}{5} = \frac{6}{10} = 60\%$ A mark-up of 60% is required.

Needless to say, it is one thing to project a profit and another to make one. The higher your mark-up, the higher your prices, and likely the lower the number of units sold. That has to be figured in, too, and the goal is to find a "sweet spot" that gets profits up without driving buyers away. The math that makes that work is beyond this book, but if you're interested, explore "mathematical optimization."

Problem 3:

An auto dealer has 100 used cars that must be sold this week. He paid $8,000 for each car and wants to cover his overhead of $20,000 and show a profit of at least $20,000. What is the minimum mark-up that he would find acceptable?

Solution:

He needs to sell the cars for at least $40,000 more than he paid for them, $20,000 for overhead and $20,000 for profit. He paid $800,000, or $8,000 × 100.

$$\frac{\$40,000}{\$800,000} = \frac{4}{80} = \frac{1}{20} = 5\%$$

Problem 4:

If he wanted to make a $60,000 profit, what mark-up would he come up with?

Solution:

For a $60,000 profit, he'd need a mark-up that would produce $80,000. $\frac{\$80,000}{\$800,000} = \frac{8}{80} = \frac{1}{10} = 10\%$

SELF-TEST 22.2

1. If you have an inventory of 200 units, which cost you $15 per unit, you have an overhead of $5,000, and you want to break even, how much is your mark-up?

2. If you have 500 dresses in your shop for which you paid $40 apiece, you have an overhead of $8,000, and you want to make a profit of $4,000, how much is your mark-up?

3. If you have an inventory of 300 units, which cost you $20 per unit, you have an overhead of $3,000, and you want to make a profit of $5,000, how much is your mark-up?

3   DISCOUNTING FROM LIST PRICE

There are certain goods that have list prices—books, appliances, cars, and sometimes furniture and clothing. These are also known as sticker prices or manufacturers' suggested retail prices. These prices are set by the manufacturer, who then sells the goods to retailers at a set discount.

Earlier we talked about a mark-up charged to the customer by the retailer. Here we have a mark-down given to the retailer by the manufacturer. Let's go over an example.

Problem 1:

If list price on a refrigerator is $500 and the manufacturer gives it to the retailer for 40% off list, how much is the retailer charged?

Solution:

You could find 40% of $500 and subtract that from $500, but you can accomplish the same thing in one step. If the retailer gets 40% off, they pay 60%.

$$\$500 \times .60 = \$300$$

So far, so good.

Problem 2:

Now the retailer turns around and charges the customer $500, right? Right! What's the percentage mark-up? (Hint: There's a trick here.)

Solution:

The retailer is marking up the refrigerator he bought for $300. The mark-up is $200. $\frac{\$200}{\$300} = \frac{2}{3} = 66\frac{2}{3}\%$

Very interesting. The manufacturer gives the retailer a 40% discount off list price (i.e., a $200 discount off the $500 list). Then the retailer raises the price by the same amount, $200, and charges the customer $500. But the percentage increase is $66\frac{2}{3}\%$ rather than 40%.

How do you explain that when the price is lowered by $200, it's a 40% decrease, but when it's raised by $200, it's a $66\frac{2}{3}\%$ increase?

The answer is that we're using different bases. When we take $200 off the $500 list price, that's $200 ÷ $500 = 40%. But when we go up from $300 to $500, we're putting $200 over a base of $300, $200 ÷ $300, and getting $66\frac{2}{3}\%$.

To sum up, when we measure percentage changes, we'll find that a given change—$200 in this case—will result in a larger percentage change when we use a smaller base—$300 instead of $500.

Now let's return to the real world. Do all retailers charge list price? Do I even have to ask? Some charge more, and some, who bill themselves as discount stores, charge less than list. Right now, however, we'll just worry about what a retailer has to pay for the merchandise.

Problem 3:

A lamp store receives a shipment of lamps, which have a list price of $29.95, at a 40% discount. What does the retailer pay for each lamp?

Solution:

$$\$29.95 \times .60 = \$17.97$$

Problem 4:

A toy store receives a shipment of 100 toys at a 20% discount from their list price of $5.95. How much does the retailer pay for this shipment?

Solution:

$$\$5.95 \times .8 = \$4.76$$

$$\$4.76 \times 100 = \$476$$

Are you getting the hang of it? I certainly hope so, because, naturally, we can't leave well enough alone. The next example gets a bit more complicated.

Problem 5:

An appliance dealer receives a shipment of 100 fans at a 30% discount off list. If she is billed $2,800, how much is the list price (of one fan)?

Solution:

Let $x =$ the list price of one fan

$$\$2,800 = 100 \times .7x$$

$$\$28 = .7x$$

$$\frac{\$28}{.7} = \frac{.7x}{.7}$$

$$\$40 = x$$

$$.7\,\overline{)\,\$28\,} = 7\,\overline{)\,\$280\,}^{\;\$40}$$

If you got this one right, go to Self-Test 22.3. If not, go to the next problem.

Problem 6:

A furniture store receives a shipment of 10 couches at a 20% discount and is billed $4,792. How much is the list price for one couch?

Solution:

Let $x =$ the list price of one couch

$$\$4{,}792 = 10 \times .8x$$

$$\$4792 = 8x$$

$$\frac{\$4792}{8} = \frac{8x}{8}$$

$$\$599 = x$$

$$\begin{array}{r} \$5\ 9\ 9 \\ 8\overline{)\,\$47^79^72} \end{array}$$

SELF-TEST 22.3

1. An audio store receives 10 turntables at a 40% discount off their list price of $99.99. How much does the retailer pay for this shipment?

2. A shoe store receives 100 pairs of shoes that list for $49.95 at a 30% discount. How much does the retailer pay for this shipment?

3. An appliance dealer receives 20 electric can openers at a 25% discount off list. If the retailer is billed $299.25, how much is the list price?

4. A video game shop receives 100 games at 35% off list. If the retailer is billed $1,946.75, how much is the list price for one game?

4   QUANTITY DISCOUNTS

One reason that "big box stores" can often sell items at a lower price than smaller stores is because they benefit from quantity discounts. Suppliers are willing to offer retailers a discount for buying a large quantity of product at one time. How large is large and how much of a discount is offered varies for different products and different suppliers. But because it's generally cheaper to make one big shipment than several smaller ones over time, most suppliers offer such quantity discounts.

Problem 1:

If a distributor offers an 8% discount on deliveries of at least 20 cases of a soft drink and 12% on deliveries of at least 100 cases, how much would a store pay for 20 cases if the regular price charged were $7 a case (for orders of under 20 cases)?

Solution:

With an 8% discount for an order of 20 cases, you pay 92%.

$$\$7 \times 20 = \$140$$

$$\$140 \times .92 = \$128.80$$

Problem 2:

How much would a store pay for an order of 150 cases?

Solution:

Because the order is over 100 cases, the discount is 12%, so you pay 88%.

$$\$7 \times 150 = \$1.050$$

$$\$1,050 \times .88 = \$924$$

Note the shortcuts we've been taking. Instead of having multiplied $1,050 by 0.12 and then subtracting that product from $1,050, we've saved ourselves a little work by multiplying by 1.00 − 0.12 = 0.88. As you keep working with figures, you'll feel comfortable taking these shortcuts too.

Obviously, it pays to buy in quantity, but quantity buying does have its costs. You need larger storage facilities, there's sometimes spoilage, there could be obsolescence (especially with fashion-oriented or computer goods), and then there is the money you have tied up in carrying larger inventories. Unless the supplier allows you to buy on credit—and doesn't offer a discount for fast payment—the carrying of a large inventory implies an interest cost for the money you've invested in that inventory.

Problem 3:

A retailer is offered a quantity discount of 5% on all orders of 50 or more and a discount of 8% on orders of 200 or more. How much would an order of 100 cost if the regular price (before the discount) were $60 per unit?

Solution:

$$\$60 \times 100 = \$6,000$$
$$\$6,000 \times 0.95 = \$5,700$$

Problem 4:

How much would the retailer have to pay for an order of 300 units?

Solution:

$$\$60 \times 300 = \$18,000$$
$$\$18,000 \times 0.92 = \$16,560$$

SELF-TEST 22.4

1. A retailer is offered a quantity discount of 6% on all orders of 100 or more and a discount of 9% on orders of 400 or more. Find how much the retailer would pay for (a) an order of 100 and (b) an order of 500. Assume a price of $5 per unit before any discounts.

2. A manufacturer offers its retail dealers the following schedule of quantity discounts: on orders over 50, a 2% discount; on orders over 200, a 4% discount; and on orders over 1,000, a 6% discount. How much would a dealer pay for an order of (a) 100, (b) 500, and (c) 2,000? Assume a price of $10 per unit before any discounts.

5   2/10 n/30

One of the biggest problems some businesses encounter is large backlogs of accounts receivable, that is, too many clients who haven't paid their bills. It's common to allow a "grace period," a time between when the bill is issued and when it is due, but suppliers will often find that their customers stretch that even further. That's a problem for the supplier, who obviously wants to get paid.

To give their business customers an incentive to pay their bills quickly, many firms offer terms of "2/10 n/30." Literally interpreted, these terms offer the buyer a 2% discount if the bill is paid within 10 days. If the buyer does not choose this option, the full amount must be paid within 30 days. It's n/30, or net/30. The word *net* refers to what the buyer owes after any other trade discounts or discounts off list. The retailer is given two options: Pay the bill within 10 days and take another 2% off, or pay the bill within 30 days and take nothing more off. For example, if you paid for a $10,000 shipment within 10 days, you would pay just $9,800, instead of paying the full $10,000 at the end of the month. Which would you do? As we've seen, interest is the price charged for the use of money over time. By offering a 2% discount for fast, or cash, payment, suppliers are really offering interest on the money. The question is, how much?

Interest is usually calculated at an annual rate. How much, then, is the 2/10 discount giving on an annual basis? What is the actual benefit, if all payments are made in the 10-day window, and receive the 2% discount?

We'll make some assumptions to simplify this question:

1. You pay in exactly 10 days to get the 2% discount.

2. You would pay on the 30th day if no discount were offered.

3. The business year consists of twelve 30-day months, or 360 days.

If you did not take the 2% discount, you would have the use of your money for an extra 20 days, but you are giving up 2% interest. How much does that come to on an annual basis?

Problem 1:

Work out the annual rate of interest. We're going to do this step by step.

Solution:

Let x = the annual rate of interest

First, we set the basic relationships: 2% interest over 20 days is equivalent to x% interest over 360 days. Write that as a proportion.

$$\frac{2\%}{20} = \frac{x\%}{360}$$

Cross-multiply: multiply the means (or inside numbers, 20 and x) and the extremes (or outside numbers, 2% and 360). The results should be equal.

$$2\%(360) = 20x$$

Do the arithmetic.

$$720\% = 20x$$

Divide by 20 to find x.

$$36\% = x$$

Incidentally, if you don't think you can find happiness with the method used to calculate the annual rate of interest, here's an alternate method.

Problem 2:

If you save 2% by paying 20 days early, how much is the daily interest rate?

Solution:

2% for 20 days is equivalent to 1% for 10 days or 0.1% for 1 day.
    The annual rate of interest is:

$$360 \times 0.1\% = 36\%$$

Which method is better? The one you're more comfortable with.

Problem 3:

What if the terms were 3/10 n/60? How much is the annual rate of interest that you would receive by paying within 10 days? (Hint: Pay within 10 days as opposed to paying how many days later?)

Solutions:

First method: Let $x =$ annual interest rate

$$\frac{3\%}{50} = \frac{x}{360}$$
$$3\%(360) = 50x$$
$$1,080\% = 50x$$
$$108\% = 5x$$
$$21.6\% = x$$

Second method:

$$3\% = 50 \text{ days}$$
$$0.3\% = 5 \text{ days}$$
$$21.6\% = 360 \text{ days}$$

How did we get from the second step to the third step? By multiplying both sides by 72, since this gave us the interest rate for 360 days.
    The key thing is to figure out that you either pay in 10 days and get the discount or pay in 60 days and don't get the discount. Therefore, if you don't pay for 60 days, you have the use of the money for an additional 50 days.

Problem 4:

What if the terms were 2/20 n/60?

Solutions:

First method: Let x = annual interest rate

$$\frac{2\%}{40} = \frac{x}{360}$$

$$720\% = 40x$$

$$72\% = 4x$$

$$18\% = x$$

Second method:

$$2\% = 40 \text{ days}$$

$$1\% = 20 \text{ days}$$

$$0.1\% = 2 \text{ days}$$

$$0.05\% = 1 \text{ day}$$

$$18\% = 360 \text{ days}$$

Does it make any sense for a business to delay paying its bills?

If you don't pay your bills promptly, you not only lose the early payer's discount, but credit agencies will list you as a slow payer, and suppliers will be very cautious about granting you credit.

But there's an advantage to slow paying as well. First of all, you might not have the money to begin with. Or, if you do have the money, the longer you delay, the longer you have the use of the money that you owe.

Problem 5:

Terms are 2/10 n/30. The firm finally pays after 4 months. If the bill is for $10,000 and the going rate of interest is 12%, how much is it worth to this retailer to have the use of $10,000 for 120 days?

Solution:

120 days = 1/3 of a year. If the annual rate of interest is 12%, then 1/3 of 12% is 4%.

$$.04 \times \$10,000 = \$400$$

It is worth $400 in what we can call implied interest for the retailer to pay 90 days late (or 120 days after being billed). So it does pay to pay late. And the higher the going rate of interest and the longer payment is delayed, the more it pays. This may offend your sense of justice, but we've got to let the bad guys win once in a while.

SELF-TEST 22.5

1. How much is the implied annual interest rate for each of the following payment terms?

   (a) 2/10 n/60

   (b) 1/30 n/90

   (c) 3/10 n/60

2. On a $1,000 bill, how much does a retailer save by paying it in 6 months if the annual rate of interest is 10%?

3. On a $100,000 bill, how much does a retailer save by paying it 3 months late if the annual rate of interest is 16%?

6   CHAIN DISCOUNTS

In this section, we're not talking about mark-downs offered by chain stores. We're talking about multiple discounts offered by suppliers to retail stores. So far, we've looked separately at three different discounts. There was the trade discount, or discount from list price. Then there was the quantity discount. And finally, there was the cash discount, which was a reward for prompt payment. A chain discount is a combination of two or three of these discounts.

To calculate a chain discount, we'll start with list price. If a manufacturer offers a trade discount of 20% off list, then the retailer would get an item listed at $100 for $80. A 5% quantity discount would knock this down to $76 ($80 × 0.95). Finally, we have 2/10 n/30. That would bring us down to $74.48 ($76 × 0.98) if we pay within 10 days.

Does it matter whether we multiply the trade discount first, then the quantity discount, and finally, the cash discount? In other words, does the order in which we multiply affect our answer? No. If you don't believe me, try it yourself.

Now we'll work out a couple of problems.

Problem 1:

Calculate how much this retailer pays, given these terms: trade discount of 25%, quantity discount of 5%, 2/10 n/30. The list price is $50.

Solution:

$$\$50 \times 0.75 \times 0.95 \times 0.98 = \$34.91$$

Problem 2:

How much does the retailer pay for an order of 500 units if the list price is $40, the trade discount is 30%, a quantity discount of 4% is offered on orders of over 100, and there are additional terms of 3/10 n/60?

Solution:

$$500 \times \$40 \times 0.7 \times 0.96 \times 0.97 = \$13,036.80$$

SELF-TEST 22.6

1. How much does the retailer pay for an order of 1,000 units if the list price is $60, the trade discount is 40%, a quantity discount of 5% is offered on orders over 500, and there are additional terms of 2/20 n/60?

2. How much does a retailer pay for an order of 500 units if the list price is $100, the trade discount is 35%, a quantity discount of 4% is offered on orders of over 100, and there are additional terms of 2/10 n/30?

3. How much does a retailer pay for an order of 100 units if the list price is $250, the trade discount is 30%, a quantity discount of 6% is offered on orders of 100 or more, and there are additional terms of 1/20 n/60?

7 PROFIT

We've finally gotten down to the bottom line: profit. Like so many other topics, profit will be calculated as a percentage. Economists keep telling us that businesspeople always try to maximize their profits. What we're concerned with here is measuring those profits.

The formula for profit is simple: sales − costs = profit

Is a profit of $100,000 good or bad? Well, that depends on a number of different considerations. How does that profit compare to your sales? How does it compare to your costs?

We need to talk about profit margin, the ratio of profit to sales. Different businesses have different profit margins. Supermarkets, for example, traditionally have extremely low profit margins. They may make just one or two cents on every dollar rung up on their cash registers. Why so few? Because we're talking about profit as a percentage of sales, and even the smallest supermarket does thousands of dollars of sales every day. Profit divided by a very large sales figure yields a small profit margin.

The formula for profits as a percentage of sales is simply $\frac{profit}{sales}$ expressed as a percent.

Now we'll plug in some numbers.

Problem 1:

A firm has sales of $10,000,000 and costs of $9,900,000. Find the profit as a percentage of sales.

Solution:

$$\text{Profit} = \text{sales} - \text{cost}$$
$$= 10,000,000 - \$9,900,000$$
$$= \$100,000$$

$$\text{Profit as a percentage of sales} = \frac{profit}{sales}$$
$$= \frac{\$100,000}{\$10,000,000}$$
$$= \frac{1}{100}$$
$$= 1\%$$

Let's do another one.

Problem 2:

A firm has sales of $200,000,000 and costs of $175,000,000. Find the profit as a percentage of sales.

Solution:

$$\text{Profit} = \text{sales} - \text{costs}$$
$$= \$200,000,000 - \$175,000,000$$
$$= \$25,000,000$$

$$\text{Profit as a percentage of sales} = \frac{\text{profit}}{\text{sales}}$$

$$= \frac{\$25,000,000}{\$200,000,000}$$

$$= \frac{25}{200}$$

$$= \frac{1}{8}$$

$$= 12.5\%$$

It's nice to know your profit as a percentage of your sales, but some business-people find it even nicer to know their profit as a percentage of their investment. How much is your investment? Well, we don't care about your initial investment. The important thing about your investment is what it's worth right now. If you went out of business today and sold your plant and equipment, inventory, your list of customers, and, if possible, your good will, how much could you get for it? That's your investment.

The formula for profit as a percentage of investment should come as no surprise: $\frac{\text{profit}}{\text{investment}}$ expressed as a percent.

Now we're ready to figure your profit as a percentage of your investment.

Problem 3:

A firm has sales of $10 million, costs of $9 million, and investments of $5 million.

Solution:

$$\text{Profit as a percentage of investment} = \frac{\$1,000,000}{\$5,000,000}$$

$$= 20\%$$

We're almost finished. Let's put all of our wisdom together in one problem.

Problem 4:

Find profit as a percentage of sales and as a percentage of investment if sales are $5 million, costs are $4.5 million, and investment is $1 million.

Solution:

$$\text{Profit} = \text{sales} - \text{cost}$$

$$= \$5,000,000 - \$4,500,000$$

$$= \$500,000$$

$$\text{Profit as a percentage of sales} = \frac{\text{profit}}{\text{sales}}$$
$$= \frac{\$500,000}{\$5,000,000}$$
$$= \frac{5}{50} = \frac{1}{10}$$
$$= 10\%$$

$$\text{Profit as a percentage of investment} = \frac{\text{profit}}{\text{investment}}$$
$$= \frac{\$500,000}{\$1,000,000}$$
$$= \frac{5}{10}$$
$$= \frac{1}{2}$$
$$= 50\%$$

SELF-TEST 22.7

1. Calculate profit as a percentage of sales:

    (a) Sales = $1 million; costs = $950,000

    (b) Sales = $500 million; costs = $200 million

2. Calculate profit as a percentage of investment:

    (a) Sales = $600,000; costs = $500,000; investment = $1 million

    (b) Sales = $50 million; costs = $45 million; investment = $20 million

3. Calculate profit as a percentage of sales and as a percentage of investment:

    (a) Sales = $1 billion; costs = $950 million; investment = $400 million

    (b) Sales = $40 million; costs = $38 million; investment = $5 million

1. Commission: $40,000 × .15 = $6,000. Total earnings for the month: $2,500 + $6,000 = $8,500.

2.
$$
\begin{array}{r}
\$5,500 \\
4,300 \\
+6,800 \\
\hline
\$16,600
\end{array}
\qquad
\begin{array}{r}
\$16,600 \\
\times .08 \\
\hline
\$1,328.00
\end{array}
$$

3. $67,500 × .10 = $6,750
$$
\begin{array}{r}
\$137,500 \\
\times .05 \\
\hline
\$6,875.00
\end{array}
\qquad
\begin{array}{r}
\$6,750 \\
+6,875 \\
\hline
\$13,625
\end{array}
$$

4. $2,500 × 12 = $30,000
$30,000 × .09 = $2,700

5.
$$
\begin{array}{r}
850000 \\
\times .025 \\
\hline
4250000 \\
17000000 \\
\hline
21250.000
\end{array}
$$

   The salesperson's share of the fee is $21,250.

1. 200 × $15 = $3,000
$$\frac{\$5,000}{\$3,000} = 167\%$$

2. $40 × 500 = $20,000
$8,000 + $4,000 = $12,000
$$\frac{\$12,000}{\$20,000} = \frac{12}{20} = \frac{6}{10} = 60\%$$

3. $20 × 300 = $6,000
$3,000 + $5,000 = $14,000
$$\frac{\$8,000}{\$6,000} = \frac{8}{6} = \frac{4}{3} = 1\frac{1}{3} = 133\frac{1}{3}\%$$

ANSWERS TO SELF-TEST 22.3

1. $999.90
    $$\frac{\times \ .60}{\$599.9400} = \$599.94$$

2. $4,995
    $$\frac{\times \ .70}{\$3,496.50}$$

3. Let x = list price of one item
    $299.25 = 20 \times .75x$
    $299.25 = 15x$
    $19.95 = x$

$$\begin{array}{r} 19.95 \\ 15\overline{)299.25} \\ {\scriptstyle X\ XX} \\ \underline{-15} \\ 149 \\ \underline{-135} \\ 14\ 2 \\ \underline{-13\ 5} \\ 75 \\ \underline{-75} \end{array}$$

4. Let x = list price of one item
    $1,946.75 = 100 \times .65x$
    $1,946.75 = 65x$
    $389.35 = 13x$
    $29.25 = x$

$$\begin{array}{r} 29.95 \\ 13\overline{)389.35} \\ {\scriptstyle X\ XX} \\ \underline{-26} \\ 129 \\ \underline{-117} \\ 12\ 3 \\ \underline{-11\ 7} \\ 65 \\ \underline{-65} \end{array}$$

ANSWERS TO SELF-TEST 22.4

1. (a) $5 \times 100 = \$500$
         $500 \times 0.94 = \$470$
    (b) $5 \times 500 = \$2,500$
         $2,500 \times 0.91 = \$2,275$

2. (a) $10 \times 100 = \$1,000$
         $1,000 \times 0.98 = \$980$
    (b) $10 \times 500 = \$5,000$
         $5,000 \times 0.96 = \$4,800$
    (c) $10 \times 2,000 = \$20,000$
         $20,000 \times 0.94 = \$18,800$

ANSWERS TO SELF-TEST 22.5

1. (a) Let x = annual interest rate

$$\frac{2\%}{50} = \frac{x}{360}$$

$$720\% = 50x$$

$$72\% = 5x$$

$$14.4\% = x$$

(b) Let x = annual interest rate

$$\frac{1\%}{60} = \frac{x}{360}$$

$$360\% = 60x$$

$$6\% = x$$

(c) Let x = annual interest rate

$$\frac{3\%}{50} = \frac{x}{360}$$

$$1{,}080\% = 50x$$

$$108\% = 5x$$

$$21.6\% = x$$

2. $.05 \times \$1{,}000 = \$50$

3. $.04 \times \$100{,}000 = \$4{,}000$

ANSWERS TO SELF-TEST 22.6

1. $\$60 \times 1{,}000 \times 0.60 \times 0.95 \times 0.98 = \$33{,}516$
2. $\$100 \times 500 \times 0.65 \times 0.96 \times 0.98 = \$30{,}576$
3. $\$250 \times 100 \times 0.70 \times 0.94 \times 0.99 = \$16{,}285.50$

ANSWERS TO SELF-TEST 22.7

1. (a) profit = sales − cost

$$= \$1{,}000{,}000 - \$950{,}000$$

$$= \$50{,}000$$

$$\text{Profit as a percentage of sales} = \frac{\text{profit}}{\text{sales}}$$

$$= \frac{\$50,000}{\$1,000,000}$$

$$= \frac{5}{100} = 5\%$$

(b)   profit = sales − cost

$$= \$500,000,000 − \$200,000,000$$

$$= \$300,000,000$$

$$\text{Profit as a percentage of sales} = \frac{\text{profit}}{\text{sales}}$$

$$= \frac{\$300,000,000}{\$500,000,000}$$

$$= \frac{3}{5} = 60\%$$

2. (a)   profit = sales − cost

$$= \$600,000 − \$500,000$$

$$= \$100,000$$

$$\text{Profit as a percentage of investment} = \frac{\text{profit}}{\text{investment}}$$

$$= \frac{\$100,000}{\$1,000,000}$$

$$= \frac{1}{10} = 10\%$$

(b)   profit = sales − cost

$$= \$50,000,000 − \$45,000,000$$

$$= \$5,000,000$$

$$\text{Profit as a percentage of investment} = \frac{\$5,000,000}{\$20,000,000}$$

$$= \frac{5}{20}$$

$$= \frac{1}{4} = 25\%$$

3. (a) profit = sales − cost

   $$= \$1,000,000,000 - \$950,000,000$$

   $$= \$50,000,000$$

   Profit as a percentage of sales $= \dfrac{\text{profit}}{\text{sales}}$

   $$= \dfrac{\$50,000,000}{\$1,000,000,000}$$

   $$= \dfrac{5}{100} = 5\%$$

   Profit as a percentage of investment $= \dfrac{\$50,000,000}{\$400,000,000}$

   $$= \dfrac{5}{40}$$

   $$= \dfrac{1}{8} = 12.5\%$$

   (b) profit = sales − cost

   $$= \$40,000,000 - \$38,000,000$$

   $$= \$2,000,000$$

   Profit as a percentage of sales $= \dfrac{\text{profit}}{\text{sales}}$

   $$= \dfrac{\$2,000,000}{\$40,000,000}$$

   $$= \dfrac{2}{40}$$

   $$= \dfrac{1}{20} = 5\%$$

   Profit as a percentage of investment $= \dfrac{\text{profit}}{\text{investment}}$

   $$= \dfrac{\$2,000,000}{\$5,000,000}$$

   $$= \dfrac{2}{5} = 40\%$$

23 A Taste of Statistics

In the 1950s, a book called *How to Lie with Statistics* by Darrell Huff got a lot of people thinking and talking about statistics and how we use them. Today we're surrounded by statistics: in public policy, science, sports, and more. It's important to ask questions when you encounter statistics. Let's start with these two questions:

1. What is Statistics? and

2. What are Statistics? (No, it's not a grammar exercise.)

Statistics is a branch of mathematics. It deals with collecting numerical data, presenting the data in understandable ways, summarizing, analyzing, and interpreting the data so that useful conclusions can be drawn. Statistics are the numerical summaries of the data collected, such as averages or percentages, that help us understand the data.

It is possible to lie about a statistic, or to lie using a statistic to support your lie. It's also possible to do a poor or flawed statistical analysis, or to present only those statistics that support your predetermined point of view. If statistics are going to be helpful to you, you'll need to understand enough about them to judge whether you should trust the statistics you read.

Because statistics are based on large collections of data—that is, lots of numbers—the calculations are generally done with calculators or computers. In this chapter, we'll talk about the calculations using small groups of numbers, so that you understand what goes on. But this chapter is more of a literacy chapter, aimed at helping you understand the statistics you encounter than actually calculating them.

1 DISTRIBUTIONS OF DATA

The first step for any statistician, before any statistics can be generated, is to collect data. There are many ways that can happen. It could be noting the results of scientific tests over the course of an experiment. It might be collecting responses of a group of people to survey questions, or it could be recording the performance of an athlete or a team over the course of a season or a career.

You may well want to ask questions about how data was collected. How was the experiment conducted? How did the experimenters decide what data to collect? How many people were interviewed? How were the people who were interviewed selected? Were the questions written in a way to avoid suggesting a particular answer? These and many more questions are reasonable inquiries before any statistics are calculated. A full exploration of these is beyond the scope of this book, but they serve as a good reminder to receive any statistic with curiosity about how it came to be.

Once data has been collected, one of the first things statisticians will do is to look at the distribution of the data. That's a first step in looking for patterns that will help make sense of a large collection of numbers.

To see the distribution, or basic pattern, of the data, researchers will often create a visual representation, a graph of some kind. This allows them to ask and answer key questions: Is there a pattern to the data, or does it look like a random scattering? What's the shape of the distribution? Is it uniform or does it have highs and lows? Does there seem to be a center or clustering point, and if so, where? How spread out is the data? Does it cluster tightly in a small area or spread out over a wide range? How high and how low does it go? Are there deviations from the pattern that might indicate an error or an unexpected result?

Suppose we present you with a data set like this:

60	60.5	61	61	61.5	63.5	63.5	63.5	64	64
64	64	64	64	64	64.5	64.5	64.5	64.5	64.5
64.5	64.5	64.5	66	66	66	66	66	66	66
66	66	66	66.5	66.5	66.5	66.5	66.5	66.5	66.5
66.5	66.5	66.5	66.5	67	67	67	67	67	67
67	67	67	67	67	67	67.5	67.5	67.5	67.5
67.5	67.5	67.5	68	68	69	69	69	69	69
69	69	69	69	69	69.5	69.5	69.5	69.5	69.5
70	70	70	70	70	70	70.5	70.5	70.5	71
71	71	72	72	72	72.5	72.5	73	73.5	74

You can identify the highest and lowest values but you probably can't tell much more before all those numbers start making your head spin. Suppose we show you the same data like this:

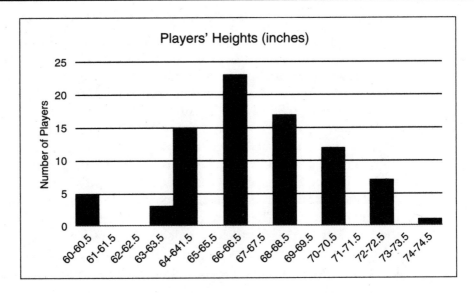

With the visual representation you can see not only the highest and lowest value, that is the tallest and shortest player, but that most of the players are between 64 and 73 inches. It looks like 66 to 66.5 inches is the most common height and you can see that the tallest bar is toward the center. You can also see that there are more players who are taller than that than there are players who are shorter. You can see that some heights were never recorded, so there are gaps between bars, and the bar for 60 to 60.5 inches seems to stand away from the others a bit, suggesting that it's an unusual occurrence.

The principal characteristics of a distribution that we want to identify are:

- **Shape:** Is it symmetric, meaning a peak in the middle and roughly the same drop off on each side, or skewed, meaning the peak is to one side and the other side trails off into a "tail"? It may not be either. It could be uniform, that is, all bars about the same, or it may not seem to fit any of these.
- **Center:** We'll calculate center in the next section, but from a look at the distribution, you can generally approximate the point the data clusters around. In a symmetric distribution, it's generally in the middle, but for skewed distributions, it will be to one side or the other. It's still the center in the sense that about half the data sits above and half below, but on one side it will be tightly packed together and on the other, spread out and trailing off into the tail.

- **Spread:** Again, we'll calculate this in a later section, but from a look at the distribution, you'll want to note whether the data is tightly clustered or more spread out.
- **Outliers and anomalies:** An outlier is a data point that stands far above or far below the bulk of the data. What "far" means has a mathematical definition, but you can generally see that value that sits off by itself.

The data on players' heights is roughly symmetric. It's not perfect, because there is more data above the peak than below, but the general appearance is peak in the middle, trailing off on both sides. The center looks to be around 66. The calculation of center will probably give a slightly higher number because there is more data above. The spread can be described by the range, the distance from the lowest to the highest value, which is $74 - 60 = 14$ inches. Other measures of spread will require a bit of calculation. The lowest bar seems to sit a bit apart from the others, but probably not enough to meet the mathematical definition of an outlier.

When the data set 4; 5; 6; 6; 6; 7; 7; 7; 7; 8 is graphed, you can see that it's not symmetric, as shown in the following image. The peak is to the right and the tail is to the left. We say this distribution is skewed, and specifically skewed to the left. The skew is always in the direction of the tail. The peak is at 7 but more data sits below that than above. In fact, there is only one data point above. The true center will get pulled down, and we might estimate it at 6. The range is $8 - 4 = 4$ units, and no outliers or other anomalies are noticeable.

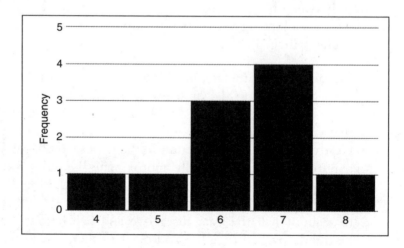

The distribution of 6; 7; 7; 7; 7; 8; 8; 8; 9; 10 is also skewed but this one is skewed to the right, as shown in the next image. The peak is at 7 but only

1 data point sits below that so the true center is probably pulled up a bit. The range is $10 - 6 = 4$ units, and no outliers appear.

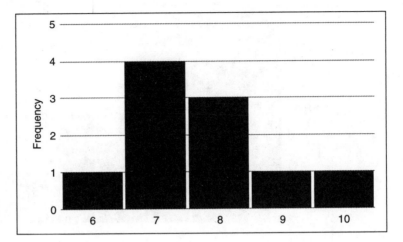

SELF-TEST 23.1

Describe the shape, center, and spread of each set of data. Note outliers or anomalies if any.

1. 6, 9, 5, 8, 9, 8, 9, 7

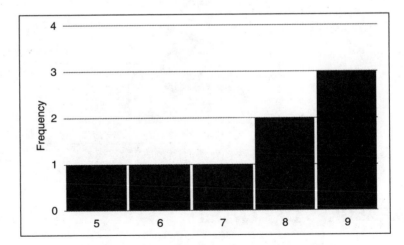

2. 1, 1, 1, 1, 2, 2, 2, 3, 3, 3, 4, 4, 5, 5, 6, 7, 8

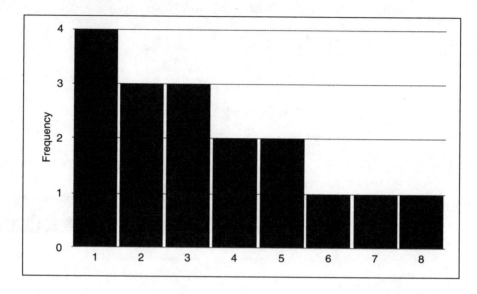

3. 2, 3, 4, 4, 5, 5, 5, 6, 6, 7, 8, 19

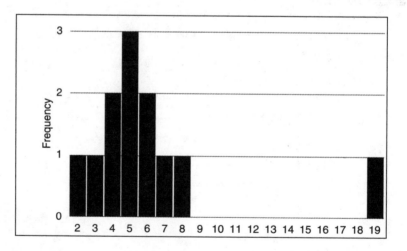

2   MEASURES OF CENTER

The mean, the median, and the mode are three very common statistical measures of center. The mean is the average of all the values in the data set. The median is the value that falls in the middle when the numbers are in order from smallest to largest or largest to smallest, and the mode is the value that occurs most often.

The mode is easy to spot, but not a reliable indicator of center, especially in a skewed distribution, or when the distribution has more than one mode. We'll focus primarily on the mean and median.

Here is a set of numbers:

$$2, 5, 7, 8, 8$$

The mode for this set of numbers is 8 because there are more 8s than any other number. But here, since 8 is also the largest value, it doesn't really describe the center.

The mean for these numbers is found by adding them all and then dividing by the number of terms: $30 \div 5 = 6$. This calculation tells us the mean of this set of numbers is 6, which may seem odd because there is no 6 in the set, but that's fine. The mean does not have to be a member of the set. Often the division will produce a decimal even though all the data points were whole numbers. Imagine a number line with these values marked off, and you're trying to balance it like a seesaw. The mean is like the fulcrum, the divider between the two sides of the seesaw. If you put it up in the $7-8$ range, the seesaw won't balance. It needs to come down to 6 so the long stretch down to 2 doesn't overwhelm the weight up at 8.

The median is another common way to describe the center. It's the number that has half of the data below it and half above it. This set of data contains five values and is already ordered from smallest to largest, so we can just pick out the third number. The median is 7, which is the middle value.

If we added another value to this set, say another 2, there would be six values, and no value absolutely in the center. In a case like that, we'd average the two values closest to center. The median of 2, 2, 5, 7, 8, 8 would be the average of 5 and 7. $5 + 7 = 12$, and $12 \div 2 = 6$.

Problem 1:

Find the mean, median, and mode for this data set:

$$6, 9, 5, 8, 9, 5, 9$$

Solution:

The mean is $51 \div 7 = 7.29$, or approximately 7.3. The median is found by putting the data set in order: 5, 5, 6, 8, 9, 9, 9. The median, or middle number, is 8. The mode is 9, which is mentioned more frequently than any of the other numbers.

What if a data set has more than one mode? What if there are *two* numbers mentioned with greater frequency than any other numbers? No problem. We'll call it a bimodal distribution.

Problem 2:

In the following data set, what are the modes?

$$8, 4, 1, 6, 3, 8, 2, 1, 7, 3, 8, 2, 1$$

Solution:

8 and 1

Problem 3:

In this data set, what is the median? What is the mean?

$$8, 10, 11, 12, 15, 16$$

Solution:

Since there is no true middle number, we should take the average of two numbers closest to the middle, 11 and 12. And what is the average of 11 and 12? We add 11 and 12 to get 23 and divide that by 2, which gives us 11.5. The median is 11.5.

The mean requires us to add all six values and divide by 6: $8 + 10 + 11 + 12 + 15 + 16 = 72$ and $72 \div 6 = 12$. The mean is 12 and the median is 11.5.

You've probably noticed that the means and medians have been similar, but not always the same. One of the reasons for this is that the mean can be affected—we say pulled—by very large or very small values. Remember problem 3 in Self-Test 23.1, the one with the outlier of 19? Without the 19, that data set would have a mean of 5. With the 19 added in, the mean jumps to 6.16.

The median, on the other hand, is resistant to very large or very small values. Imagine you listed the salaries of all the people in a company, in order, smallest to largest. (The largest is the CEO's salary.) You find the median, or middle value. And then the CEO doubles his own salary. Does the middle move?

SELF-TEST 23.2

1. Find the mean, median, and mode of this data set: 31, 16, 22, 19, 17, 22, 15

2. Find the mean, median, and mode of this data set: 27, 14, 2, 9, 4, 13, 14, 7

3. Find the mean, median, and mode of this data set: 4, 5, 9, 4, 8, 1, 10, 5, 1, 3

3 MEASURES OF SPREAD

Earlier we used the idea of range, the distance between the highest and lowest value, to describe the spread of a distribution. Range is easy to calculate, and easy to use for comparing two distributions. The distribution with the bigger range is more spread out. But there's a lot of information that the range doesn't communicate. Is the data all clumped up at one end of the range with just a scattering on the other end? Is a big range caused by one outlier? How evenly or unevenly is the data spread over that range?

To get a better picture, there are other ways to describe spread, some simple to find, one more complicated. We'll look at each one, and when they're usually used.

Median and Quartiles

In Frame 2, we talked about the median, or middle value, as a measure of center. When we use the median as our center it's common to use quartiles to help measure spread.

The median is the value that divides the data into two groups of equal size. The quartiles are the values that divide the data into four groups of equal size. The first quartile, or Q1, separates the lowest 25% of the data from the rest. You rarely see any mention of a second quartile or Q2 because that is actually the median. The third quartile, or Q3, is the value that separates the highest 25% of the data from the lower values.

Another, perhaps simpler, way to think about quartiles is that the median divides the data set into halves. Q1 is the median of the lower half. Q3 is the median of the upper half.

The toughest part of finding a median is getting the data in order from smallest to largest. If you've already found the median, that's done, and finding the quartiles is just a matter of locating the middle of each section.

Problem 1:

Find the median and quartiles of this data set: 4, 7, 9, 6, 1, 5, 12, 4, 15, 11, 3, 8, 2, 13

Solution:

Put the data in order from lowest to highest: 1, 2, 3, 4, 4, 5, 6, 7, 8, 9, 11, 12, 13, 15

There are 14 values; we'll have to average the two in the middle to find the median. $6 + 7 = 13$ and $13 \div 2 = 6.5$. The median is 6.5.

To find Q1, find the median of 1, 2, 3, 4, 4, 5, 6. Q1 = 4

To find Q3, find the median of 7, 8, 9, 11, 12, 13, 15. Q3 = 11

Problem 2:

Find the median and quartiles of this data set: 29, 35, 18, 41, 32, 38, 27, 34, 40, 35, 51

Solution:

Put the data in order from lowest to highest: 18, 27, 29, 32, 34, 35, 35, 38, 40, 41, 51

There are 11 values, so the median is the middle value, 35.
To find Q1, find the median of 18, 27, 29, 32, 34. Q1 = 29
To find Q3, find the median of 35, 38, 40, 41, 51. Q3 = 40

5-Number Summary

Once we've found the median and the quartiles, we can describe the distribution using what's known as the 5-number summary: the minimum value, the first quartile, the median, the third quartile, and the maximum value.

5-number summary: MIN – Q1 – MED – Q3 – MAX

For the data set 2, 3, 8, 15, 12, 7, 8, the minimum value is 2 and the maximum is 15. If we order the data, we can see that the median is 8, Q1 is 3, and Q3 is 12. The 5-number summary is 2 – 3 – 8 – 12 – 15. From the 5-number summary, we can see that the center of the distribution is at 8, and half the data falls between 3 and 12, with the median a little closer to 12. The max and min are not far from the quartiles, so there are no outliers.

You've already done the work for these two problems, so solving them should be quick.

Problem 3:

Give the 5-number summary for 4, 7, 9, 6, 1, 5, 12, 4, 15, 11, 3, 8, 2, 13 (from Problem 1).

Solution:

$$1 - 4 - 6.5 - 11 - 15$$

Problem 4:

Give the 5-number summary for 29, 35, 18, 41, 32, 38, 27, 34, 40, 35, 51 (from Problem 2).

Solution:

$$18 - 29 - 35 - 40 - 51$$

The 5-number summary is sometimes represented visually by a boxplot (or box-and-whisker plot.) The box, or rectangle, stretches from Q1 to Q3, with a vertical mark at the median. The "whiskers" at either side of the rectangle stretch out to the minimum and maximum values.

Here's the boxplot for Problem 3:

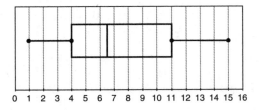

And here's the boxplot for Problem 4:

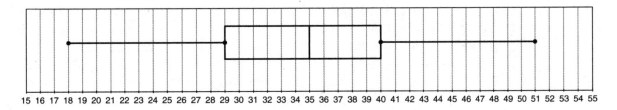

Range and Interquartile Range

We saw early in this chapter that the range is the difference between the maximum value in the data set and the minimum value. The name should tell you that the interquartile range is the distance between the quartiles, or Q3 − Q1. If you're still looking at the boxplots above, the interquartile range, or IQR, is the length of the box.

The IQR gives us a way to define outliers more precisely. An outlier is any value that falls more than 1.5 times the IQR above Q3 or below Q1. The data set from Problem 1 and Problem 3 has a 5-number summary of 1 − 4 − 6.5 − 11 − 15. To determine if it contains any outliers, we first find the IQR = 11 − 4 = 7, and then multiply that by 1.5. 1.5 × IQR = 1.5 × 7 = 10.5. Any values more than 10.5 units above 11, that is, 21.5 or higher, would be outliers. The largest value is 15, so there are no outliers on the top end. Are there any values below Q1 − 10.5? If we subtract 10.5 from 4, we get −6.5, and there are no negative numbers in the data set, so there are no outliers.

Problem 5:

Find the IQR of 29, 35, 18, 41, 32, 38, 27, 34, 40, 35, 51 (from Problem 2 and Problem 4) and determine if there are any outliers.

Solution:

The IQR = 40 − 29 = 11 and 1.5 × 11 = 16.5. Adding 16.5 to Q3 gives 40 + 16.5 = 56.5 and subtracting 16.5 from Q1 gives 29 − 16.5 = 12.5. The largest value in the set is 51 and the smallest is 18, so there are no outliers in this data set.

Mean and Standard Deviation

When we describe center with the mean, we usually describe the spread with the standard deviation. We'll tell you right now that for a data set of any size, calculating the standard deviation is a multistep process with math that can be tedious and sometimes difficult. Even for the small data sets we're going to look at, you'll probably want a calculator at least for some of the steps. For large data sets, most people trust the job to a computer or sophisticated calculator.

But it is possible to find a standard deviation mostly by hand, and we'll do that for a small data set, so you can understand what's going on and why. Here are the steps and the reason for them:

1. **Find the mean.**

    The mean is the center of the distribution. The standard deviation is a way to measure how far from that center values fall.

2. **List all the data values.**

    We're going to do the same calculation for each data value. The more organized you can be, the fewer mistakes will be made. It may help to list the data in order from smallest to largest or largest to smallest.

3. **Subtract the mean from each data value.**

    We want to know how far from the center each value falls. Notice that this subtraction will produce negative numbers for values below the mean and positive numbers for values above the mean.

4. **Square the result of each subtraction.**

    You might start looking for your calculator here if the subtraction gave you numbers that are hard to square. Why are we squaring them? Because some of the distances from the mean are positive and some are negative; if we just average them as they are, the positive and negative numbers could cancel each other out and tell us there's no spread, when we can clearly see that there is. So we square the distances, and the squares are all positive. We'll undo the squaring at the end.

5. **Add up all the squares and divide the total by the number of data values.**

    This gives us the average of the squared deviations.

6. **Take the square root of the result.**

    This is where we undo that squaring that we did earlier. Now if, in step 5, you got an average of 9 or 25, taking the square root won't be a problem. If you got 13 or 77 or 235, you'll probably want a calculator for this step.

Problem 6:

Earlier in this chapter, we found the mean of the data set 8, 10, 11, 12, 15, 16 to be 12. Let's find the standard deviation.

Solution:

We've already done step 1, finding the mean, which is 12. We'll use a table to organize steps 2, 3, and 4.

Step 2: Value	8	10	11	12	15	16
Step 3: Value – Mean	−4	−2	−1	0	3	4
Step 4: (Value – Mean), Squared	16	4	1	0	9	16

5. Add $16 + 4 + 1 + 0 + 9 + 16 = 46$ and divide $46 \div 6 \approx 7.7$

6. A reasonable time to use a calculator: $\sqrt{7.7} \approx 2.7688$. We can round that to 2.77.

SELF-TEST 23.3

1. Find the 5-number summary of this data set: 16, 22, 19, 17, 22, 15, 11, 25.

2. Find the interquartile range of this data set: 27, 14, 2, 9, 4, 13, 14, 7. Does this distribution have any outliers?

3. Find mean and standard deviation of this data set: 4, 5, 9, 4, 8, 1, 10, 5, 1, 3.

4 ESTIMATING

Statistics are used in many areas and in different ways. Many times, the statistics simply describe what happened. If you follow any professional sport, you probably have seen a great many statistics, recorded over many years, that are meant to tell you what players or teams did. You may use that information to decide, or at least argue about, who was the best at this or that aspect of the sport. But basically, those statistics just describe what happened.

The other principal use of statistics is to estimate the value of some factor that can't be directly calculated. If you wanted to investigate the average salary for public school teachers in the United States, you couldn't realistically collect salary information from every single U.S. public school teacher to average out. But you could get that information from a subgroup and use the average you got from that group to estimate the average for all teachers.

All public school teachers in the United States make up the *population*. The subgroup from which you collect data is called a *sample*. The average salary for the population is called a *parameter*, and the average for the sample is a *statistic* that can be used to estimate the parameter. (Notice we say "estimate" and not "predict." This isn't fortune telling.)

You could spend a great deal of time learning all the ins and outs of doing this sort of estimating properly, but we'll only give you a short version. We just want to help you think critically about the estimates you see.

To be successful, the process needs to begin with a clear definition of the population and the parameter. Who or what are you interested in and exactly what do you want to know? Are you interested in all teachers in the United States or just those in your state? Public school, private school, or both? The more clearly you define your population the smoother things will go. Obviously, you need to know what information you're looking for. You don't collect shoe sizes if you're interested in income.

The sample needs to be a good representation of the population. If you're interested in teacher salaries you don't choose a sample of nurses or bankers or bricklayers. The subgroup should be chosen by a simple random sample, which is a short way of saying that every member of the population has an equal chance of being in the sample. You don't just call your friends, or interview people walking out of the local high school. There are other requirements to think about but we're not actually going to choose a sample, so we'll skip to how the process of estimation works.

If you choose a sample of 100 teachers, collect salary information from those teachers and average that, you get a statistic, specifically a sample statistic: the average salary of these 100 teachers. If you repeated the process, chose a difference sample of 100 teachers, and found their average salary, it might happen to be the same as the first sample or it might be different. But what statisticians know, from experience as well as theory, is that if you did that over and over and over again, and you treated all the averages as your data, that data would have a distribution clustered pretty tightly around a center that is the actual mean of the population. No one sample is guaranteed to give you the average for the population, but the pattern of all the sample averages will point to the actual average salary of the whole population.

If we know the mean of the sample, we know that it is probably close to the actual mean of the population. If we had done hundreds of samples and had all the averages, they'd be tightly packed around the real population average, so the one we have is likely to be in the neighborhood. And they're so tightly packed that the spread of all those averages is a fraction of the standard deviation of your sample.

To estimate the population mean, we start with the sample mean, and go up a little and down a little. How far up and how far down we go depends the standard deviation, the size of the sample, and how confident we want to be that the true population parameter is between those limits. If you're content with a 50-50 chance that we'll catch the true average salary for the population, we don't need to go very far from the sample mean. If we want to be 90% certain, we need to stretch more to have a better chance to catch the true value. The higher the confidence you want to have, the farther apart we're going to place the upper and lower limits. Think of it this way: the more certain we want to be that we'll catch the true parameter, the bigger net we want to use.

What we're talking about here is building a *confidence interval* that is likely to contain the true average teacher salary for our population. Luckily there's a formula. If the sample size is n, the sample mean is $\bar{x}$, and the sample standard deviation is s, then the confidence interval is

$$\bar{x} \pm z^* \frac{s}{\sqrt{n}}$$

$z^\star$ is a value that depends on the confidence level you want. The higher the confidence level, the larger $z^\star$ will be. The following table gives $z^\star$ values for common confidence levels.

Confidence level	90%	95%	98%	99%
z*	1.645	1.960	2.326	2.576

The part of that confidence interval formula that comes after the ± sign, the amount we add to the sample mean and subtract from the sample mean, is called the *margin of error*.

Problem 1:

A random sample of 100 public school teachers in the United States had a mean salary of $52,000 per year and a standard deviation of $15,000. Find a 90% confidence interval for the average annual salary of public school teachers in the United States.

Solution:

$\bar{x}$= \$52,000, s = \$15,000, n = 100, $z^\star$ = 1.645

$$\bar{x} \pm z^* \frac{s}{\sqrt{n}} = \$52,000 \pm 1.645 \times \frac{\$15,000}{\sqrt{100}}$$

$$= \$52,000 \pm 1.645 \times \frac{\$15,000}{10}$$

$$= \$52,000 \pm 1.645 \times \$1,500$$

$$= \$52,000 \pm \$2,467.50$$

We can say we are 90% confident that the true average salary for all U.S. public school teachers is between \$49,532.50 and \$54,467.50.

Problem 2:

A researcher is interested in prices for a gallon of whole milk in supermarkets around her state. She collects prices from 900 supermarkets selected at random and calculates that the mean of this sample is \$3.59 with a standard deviation of \$0.37. If she wants to develop a 98% confidence interval for the population parameter, what margin of error should she use?

Solution:

$\bar{x}$ = \$3.59, s = \$0.37, n = 900, z* = 2.326

$$\bar{x} \pm z^* \times \frac{s}{\sqrt{n}} = \$3.59 \pm 2.326 \times \frac{\$0.37}{\sqrt{90}}$$

$$= \$3.59 \pm 2.326 \times \frac{\$0.37}{30}$$

$$= \$3.59 \pm 2.326 \times \$0.0123$$

$$= \$3.59 \pm \$0.0286$$

The margin of error will be \$0.0286, which we can round to \$0.03. The population parameter will be in a 6-cent range from \$3.56 to \$3.62.

When statisticians estimate a proportion or percentage, rather than a mean, the process is similar with slight differences in the confidence interval formula. When you see polling reports saying the such-and-such a percentage of the people surveyed said this or that, remember to look at the fine print for the margin of error.

Problem 3:

Polling in the race for country executive says Christopher Smith leads Christine Smyth. Mr. Smith has 46% of the vote and Ms. Smyth has 44%. If the margin of error is 3%, is Smith's victory guaranteed?

Solution:

Mr. Smith's share of the vote is estimated to be 46% ± 3% or between 43% and 49%. Ms. Smyth's share is estimated to be 44% ± 3% or between 41% and 47%. Neither candidate is assured of victory. Polling results like these are sometimes described as "within the margin of error" or "a statistical tie."

SELF-TEST 23.4

Use of a calculator may be appropriate.

1. A sample of metal is weighed 9 times. The mean of the weights recorded is 5.025 grams with a standard deviation of 0.006 grams. Find a 99% confidence interval for the actual weight of the metal.

2. A group of 36 students are selected at random from a large university lecture. Their grades on an exam are collected and the mean score is found to be 76 points, with a standard deviation of 5.67 points. Find a 95% confidence interval for the average score on the exam.

3. In October 2019, a poll showed that 51% of respondents favored a national health plan while 47% opposed it. Can you say with confidence that more people favor such a plan than oppose it if the margin of error is 4%?

ANSWERS TO SELF-TEST 23.1

1. Skewed left, center not at the peak, probably around 8, range 9 − 5 = 4.
2. Skewed right, center not at the peak, probably around 3, range 8 − 1 = 7.
3. Symmetric, center approximately 5, range 19 - 2 = 17, outlier at 19.

ANSWERS TO SELF-TEST 23.2

1. Mean = 142 ÷ 7 = 20.3; median = 19; mode = 22
2. Mean = 90 ÷ 8 = 11.3; median = 22/2 = 11; mode = 14
3. Mean = 50 ÷ 10 = 5; median = 9/2 = 4.5; modes = 1, 4, 5

1. 11, 15, 16, 17, 19, 22, 22, 25. Median: $(17 + 19) \div 2 = 18$, Q1 = 15.5, Q3 = 22, Min = 11, Max = 25. 5-number summary: 11 – 15.5 – 18 – 22 – 25

2. 2, 4, 7, 9, 13, 14, 14, 27. Med = 11, Q1 = 5.5, Q3 = 14, IQR = 8.5. Q1 – 1.5 × IQR will be a negative value. Q3 + 1.5 × IQR = 14 + 1.5 × 8.5 = 14 + 12.75 = 26.75 so 27 is (just barely) an outlier.

3.

Values	1	1	3	4	4	5	5	8	9	10	Mean = 50 ÷ 10 = 5
Value – Mean	−4	−4	−2	−1	−1	0	0	3	4	5	
(Value – Mean) Squared	16	16	4	1	1	0	0	9	16	25	Total of Squares = 88

$$\text{Mean} = 5, \text{ Standard Deviation} = \sqrt{\frac{88}{10}} = \sqrt{8.8} \approx 2.966$$

1. $\bar{x} = 5.025$, $s = 0.006$, $n = 9$, $z^* = 2.576$

$$\bar{x} \pm z^* \times \frac{s}{\sqrt{n}} = 5.025 \pm 2.576 \times \frac{0.006}{\sqrt{9}}$$

$$= 5.025 \pm 2.576 \times \frac{0.006}{3}$$

$$= 5.025 \pm 2.576 \times 0.002$$

$$\approx 5.025 \pm 0.0052$$

The 99% confidence interval is from 5.0198 to 5.0302.

2. $\bar{x} = 76$, $s = 5.67$, $n = 36$, $z^* = 1.960$

$$\bar{x} \pm z^* \times \frac{s}{\sqrt{n}} = 76 \pm 1.960 \times \frac{5.67}{\sqrt{36}}$$

$$= 76 \pm 1.960 \times \frac{5.67}{6}$$

$$= 76 \pm 1.960 \times 0.93$$

$$\approx 76 \pm 1.823$$

The 95% confidence interval is from 74.177 points to 77.823 points.

3. With a 4% margin of error, the portion of people who favor the plan is estimated to be between 51% – 4% = 47% and 51% + 4% = 55%. The proportion of people who oppose the plan is estimated to be between 47% – 4% = 43% and 47% + 4% = 51%. These estimates do not give a clear decision.

24 Review

Congratulations! You've just begun the last chapter of this book. Absolutely no new material will be introduced here. This chapter is composed of questions drawn from each of the earlier chapters starting with Chapter 3. It's kind of a final exam, except that it covers material from maybe 10 years of math rather than from just a term's work.

Almost no one gets 100 on a final exam, so we would be amazed if you got everything right. If you do get a particular problem wrong, you'll have the option of going back to the frames where that problem was introduced and gone over.

Again, don't worry if you get some of these wrong. If you've gotten this far, you've already absorbed about 10 years of math. And if you don't want to go back over each of the problems you get wrong, no one is looking over your shoulder.

CHAPTER 3 REVIEW

1.
$$\begin{array}{r} 803 \\ \times\, 576 \end{array}$$

2.
$$\begin{array}{r} 3195 \\ \times\, 8917 \end{array}$$

3. There are 1,760 yards in a mile. If you drove 67 miles, how many yards did you cover?

4. A car dealer sold 84 cars at $17,095 each. What were her total sales?

CHAPTER 4 REVIEW

1. $8\overline{)905}$

2. $22\overline{)1371}$

3. $752\overline{)69323}$

4. At a weight reduction center, 135 people lost a total of 20,925 pounds in 2 years. If each person lost exactly the same amount, how many pounds did each person lose?

5. If a path is 756 inches long, how many yards long is that path?

CHAPTER 5 REVIEW

1. Multiply each of these numbers by 10:

(a) .008   (b) .9   (c) 401

2. Multiply each of these numbers by 100:

(a) 700   (b) 2.1   (c) .007

3. Divide each of these numbers by 10:

(a) 0.01   (b) 1.3   (c) 618

4. Divide each of these numbers by 100:

(a) 0.03   (b) 67   (c) 4,005

5. Use mental math tricks to perform each multiplication:

(a) $92 \times 7,000$ (c) 17×22 (e) 47×5

(b) 125×303 (d) 83×11 (f) 86×500

CHAPTER 6 REVIEW

1. Add -4 and -5.   **4.** Multiply 5 and -2.

2. Add -3, $+4$, -1, and -2.   **5.** Divide -8 by -2.

3. Multiply -2 and -3.   **6.** Divide 9 by -3.

CHAPTER 7 REVIEW

1. Add these fractions:

(a) $\dfrac{1}{3} + \dfrac{1}{2} + \dfrac{3}{4}$ (b) $\dfrac{1}{5} + \dfrac{1}{2} + \dfrac{2}{3}$

2. Subtract these fractions:

(a) $\dfrac{2}{5} - \dfrac{1}{3}$ (b) $\dfrac{3}{4} - \dfrac{2}{3}$

3. Multiply these fractions:

(a) $\frac{1}{4} \times \frac{3}{5}$

(b) $\frac{2}{3} \times \frac{2}{5}$

4. Find one-eighth divided by one-fourth.

5. Find three-eighths divided by one-seventh.

6. Subtract $\frac{4}{21} - \frac{2}{15}$

7. Add $8\frac{3}{5} + 9\frac{7}{8}$

CHAPTER 8 REVIEW

1. $\begin{array}{r} 1.09 \\ \times\ 8.6 \end{array}$ 2. $\begin{array}{r} 14.562 \\ \times\ 4.225 \end{array}$ 3. $8.96\overline{)3.57}$ 4. $1.53\overline{)4.02}$

5. Stacy received payments of $38.53, $42.19, and $85.12. From the money she received, she paid a bill for $22.95. How much money is left?

CHAPTER 9 REVIEW

1. Convert $\frac{5}{8}$ into a decimal.

2. Convert $\frac{3}{4}$ into a decimal.

3. Convert 0.6 into a fraction.

4. Convert 0.45 into a fraction.

5. Convert $\frac{2}{9}$ to a decimal.

6. Convert 0.272727 … to a fraction.

CHAPTER 10 REVIEW

1. What is the ratio of inches per foot?

2. If an employee was out sick on 4 of 72 work days, what is her ratio of sick days to days worked?

3. Solve for x: 8 is to 5 as 40 is to what?

4. x:5 = 12:20. Find x.

5. If you can drive 300 miles in 6¼ hours, how long would it take you to drive 700 miles at the same rate?

CHAPTER 11 REVIEW

1. If you walked 17 miles at an average speed of 14¼ minutes per mile, how long did it take you to walk the entire distance?

2. How much is three-fifths of two-thirds?

3. If silver is selling at $6.25 an ounce, how much silver could you buy for $580?

CHAPTER 12 REVIEW

1. What is the area of a rectangle that is 8 feet long and 5 feet wide?

2. What is the area of a square that has a side of 5 feet?

3. If land is selling for $200 a square foot, how much would you have to pay for a rectangular lot that is 200 feet by 40 feet?

4. If a room were 20 feet by 45 feet and carpeting cost $20 per yard, how much would it cost for wall-to-wall carpeting?

5. How much would it cost to put a fence around a field that is 40 feet by 150 feet if fencing cost $15 a foot?

6. Find the perimeter of a square lot whose side is 20 feet.

7. If a triangle has a base of 10 inches and is 6 inches high, what is its area?

CHAPTER 13 REVIEW

1. If the diameter of a circle is 10 inches, how much is its circumference?

2. If the circumference of a circle is 15 feet, how much is its diameter?

3. If the radius of a circle is 5 inches, how much is its area?

4. If the diameter of a circle is 10 feet, how much is its area?

CHAPTER 14 REVIEW

1. Convert $\dfrac{17}{100}$ into a percent.

2. Convert .82 into a percent.

3. Convert $\dfrac{5}{8}$ into a percent.

4. If you grew from 5 feet 4 inches to 5 feet 9 inches, by what percent did your height increase?

5. A change from 15 to 60 represents a percentage change of how much?

6. If a high school has 100 freshmen, 90 sophomores, 80 juniors, and 70 seniors, find the percentage share of the student body of each class.

7. If you left a 15% tip, how much would you leave on a restaurant check of $44.80?

CHAPTER 15 REVIEW

1. Find x:

(a) $x + 4 = 9$ (b) $x - 3 = 8$

2. Find x:

(a) $3x = 20$ (b) $\dfrac{x}{5} = 1$

3. Find x:

(a) $.2x = 3$ (b) $1.5x = 2$

4. Find x:

(a) $x + 2 = -5$ (b) $x + 4 = 3$

5. Find x:

(a) $5x - 2 = 8$ (b) $\dfrac{3}{4}x + 5 = 8$ (c) $4.2x + 7 = 1$

CHAPTER 16 REVIEW

1. If $x = 4$, how much is x^1?

2. If $x = 5$, how much is x^3?

3. If $x = 2$, how much is x^4?

4. What is the square root of 81?

5. What is the square root of 49?

6. How much is x, if $x^2 = 16$?

7. How much is x, if $x^2 = 100$?

8. How much is x, if $2x^2 - 2 = 16$?

9. How much is x, if $3x^2 + 5 = 53$?

10. If $x = 9$, how much is $160 - x^2 + 3x$?

11. If $x = 8$, how much is $4x^2 + 3x - 190$?

CHAPTER 17 REVIEW

1. Translate these words into numbers:

(a) seven hundred thirty-six thousand, two hundred

(b) forty-three million

2. Please express these numbers in words:

(a) 987,100,000   (b) 189,000

3. Translate these words into numbers:

(a) four trillion   (b) two billion

4. Please express these numbers in words:

(a) 7,000,000,000,000   (b) 983,000,000,000

5. How much is $100,000 \times 87,661$?

6. How much is $4,130 \times 200,000$?

7. Divide 890 million by 1,000.

8. Divide 450 billion by 10,000.

9. Write in standard form:

(a) 8.63×10^8 (b) 9.15×10^{-4}

10. Multiply: $8 \times 10^7 \times 9 \times 10^3 \times 4 \times 10^{-12}$

11. Divide:

(a) 8.46×10^{15} by 4×10^{12} (b) 9.63×10^{-11} by 3×10^{-9}

CHAPTER 18 REVIEW

1. If Janice is 3 years short of being four times Arlene's age, and the sum of their ages is 97, how old are they?

2. Ron is five times Jason's age. In 5 years he will be just three times as old as Jason. How old are they?

3. Find three consecutive numbers adding up to 51.

4. Find three numbers adding up to 72, if the second number is three times as large as the first and the third is two larger than the second.

5. A 50-pound mixture of peanuts and cashews sells for $6 a pound. If peanuts sell for $2 a pound and cashews sell for $8 a pound, how many pounds of peanuts and how many pounds of cashews are used in the mixture?

CHAPTER 19 REVIEW

1. Calculate the simple rate of interest paid on $1,000 for 6 months at an annual rate of 6%.

2. Calculate the simple rate of interest paid on $2,000 for 2 years at an annual rate of 7%.

3. Ten thousand dollars is put in a bank that pays interest at an annual rate of 12%, compounded quarterly. How much money would be in the account after 3 quarters?

4. Four thousand dollars is lent out at an annual rate of 8% for 1 year. If the interest is compounded quarterly, how much money would the borrower owe the lender after 1 year?

5. How much is the doubling time for a compound annual interest rate of 7%?

6. How much is the doubling time for a compound annual interest rate of 1%?

A Word on Formulas

Can you show where it is written that you must memorize a lot of formulas? In this book, we happen to have covered a few fairly complex formulas. We give you our word that we will not think badly of you if you look up these and other formulas.

Formulas are tools to be used, not memorized. The only thing you need to memorize is the multiplication table up to 10 × 10.

CHAPTER 20 REVIEW

1. If a plane took off at noon and flew at 500 m.p.h. until 4:30 P.M., how far did it fly?

2. Joan left work at 5 P.M. and walked 4 m.p.h. until 6:30 P.M. and then took a bus the rest of the way home. How far does she live from work if the bus traveled at 10 m.p.h. and left her at her door at 7:00 P.M.?

3. If a plane went 3,000 miles in 5½ hours, what was its average rate of speed?

4. If Mike drove from home to work, a distance of 40 miles, in 45 minutes and returned by a different route in 1 hour, what was his average rate of speed?

5. Two trains left the station traveling in opposite directions. One train was traveling at the rate of 60 m.p.h., and the other was traveling at the rate of 55 m.p.h. When they were 460 miles apart, for how long had they been traveling?

6. Marcia walked at an average rate of speed of 3½ m.p.h. and covered a distance of 16 miles. How long did she walk?

CHAPTER 21 REVIEW

1. A living room set is marked down from $969 to $629. By what percent has the price been cut?

2. An auto dealer is offering a $1,500 rebate on a $20,000 car. What percentage of the original price do you get back?

3. If the sales tax on a $600 purchase is 5%, how much tax would you pay?

4. How much would the original price be if you paid a total (including taxes) of $428 for a sofa and the sales tax were 7%?

5. How much interest would you pay on a credit card balance of $2,500 in 1 month if the annual rate of interest were 16%?

6. If you owe $50 interest on a monthly credit card balance of $4,000, what are the annual and monthly interest rates that you must pay?

CHAPTER 22 REVIEW

1. A 10% commission is earned on sales to new customers, and a 4% commission is paid on sales to regular customers. A saleswoman wrote up sales of $74,300 to new customers and $110,000 to regular customers. How much did she earn in commissions?

2. A real estate agency charges prospective tenants 15% of their annual rent to find them apartments. If that fee is split evenly between the agency and the salesperson, how much does the salesperson make if she finds someone a $900/month apartment?

3. You have an inventory of 100 units, which cost you $80 per unit, and have an overhead of $4,000. If you want to make a profit of $2,000, how much is your mark-up?

4. A shoe store receives 100 pairs of shoes, which list for $59.95, at a 40% discount. How much does the owner pay for this shipment?

5. A manufacturer offers its retail dealers the following schedule of quantity discounts: on orders over 100, a 3% discount; on orders over 500, a 5% discount; and on orders over 1,000, a 7% discount. If the list price is $20, how much would a dealer pay for the following orders?

   (a) 150   (b) 1,500

6. How much is the implied annual interest rate for these two terms?

   (a) 2/10 n/30   (b) 3/20 n/60

7. How much does a retailer pay for an order of 1,000 units if the list price is $40, the trade discount is 40%, a quantity discount of 5% is offered on orders of 500 or more, and there are additional terms of 2/10 n/60?

8. Calculate profit as a percentage of sales and investment when sales = $1 billion, cost = $900 million, and investment is $400 million.

CHAPTER 23 REVIEW

1. Find the mean, median, and mode of this data set: 19, 1, 8, 1, 5, 6, 9.

2. Find the mean, median, and modes of this data set: 2, 23, 7, 12, 6, 2, 7, 5.

3. Find the five-number summary for this data set: 71, 62, 71, 70, 91, 65, 64, 60, 99, 60.

4. What is the IQR (interquartile range) for the data set in question 3?

5. Does the data set in question 3 contain any outliers? Explain.

6. If polling with a margin of error of 3% shows a candidate has the support of 46% of voters, in what range would you expect the candidate's final percentage of the vote to fall?

ANSWERS FOR CHAPTER 3 REVIEW

1.
$$
\begin{array}{r}
803 \\
\times\,576 \\
\hline
4818 \\
5621 \\
4015 \\
\hline
462528
\end{array}
$$

2.
$$
\begin{array}{r}
3195 \\
\times\,8917 \\
\hline
22365 \\
3195 \\
28755 \\
25560 \\
\hline
28489815
\end{array}
$$

3.
$$
\begin{array}{r}
1760 \\
\times\,67 \\
\hline
12320 \\
10560 \\
\hline
117920
\end{array}
$$
  yards

4.
$$
\begin{array}{r}
\$17095 \\
\times\,84 \\
\hline
68380 \\
136760 \\
\hline
\$1{,}435{,}980
\end{array}
$$

ANSWERS FOR CHAPTER 4 REVIEW

1.
$$
\begin{array}{r}
1\,1\,3.1\,2\,5 = 113.1 \\
8\,)\,\overline{9^{1}0^{2}5.^{1}0^{2}0^{4}0}
\end{array}
$$

2.
$$
\begin{array}{r}
62.3 \\
22\,)\,\overline{1371.0} \\
{}^{\mathrm{xx}} \\
-132 \\
\hline
51 \\
-44 \\
\hline
7\,0 \\
-6\,6 \\
\hline
\end{array}
$$

3.
$$
\begin{array}{r}
92.18 = 92.2 \\
752\,)\,\overline{69323.00} \\
{}^{\mathrm{x}} \\
-6768 \\
\hline
1643 \\
-1504 \\
\hline
139\,0 \\
-75\,2 \\
\hline
63\,80 \\
-60\,16 \\
\hline
\end{array}
$$

4.
$$
\begin{array}{r}
155 \text{ pounds} \\
135\,)\,\overline{20925} \\
{}^{\mathrm{xx}} \\
-135 \\
\hline
742 \\
-675 \\
\hline
675 \\
-675 \\
\hline
\end{array}
$$

5.
$$
\begin{array}{r}
21 \text{ yards} \\
36\,)\,\overline{756} \\
{}^{\mathrm{x}} \\
-72 \\
\hline
36 \\
\end{array}
$$

ANSWERS FOR CHAPTER 5 REVIEW

1. (a) 0.08

(b) 9

(c) 4,010

2. (a) 70,000

(b) 210

(c) 0.7

3. (a) 0.001

(b) 0.13

(c) 61.8

4. (a) 0.0003

(b) 0.67

(c) 40.05

5. (a) $92 \times 7{,}000 = 92 \times 7 \times 1{,}000 = 644 \times 1{,}000 = 644{,}000$

(b) $125 \times 303 = 125 \times 300 + 125 \times 3 = 37{,}500 + 375 = 37{,}875$

(c) $17 \times 22 = 17 \times 2 \times 11 = 34 \times 11 = 374$

(d) $83 \times 11 = 8^{8+3}3 = \overset{1}{8}13 = 913$

(e) $47 \times 5 = 47 \times 10 \div 2 = 470 \div 2 = 235$

(f) $86 \times 500 = 86 \times 5 \times 100 = 86 \times 10 \div 2 \times 100 = 860 \div 2 \times 100 = 430 \times 100 = 43{,}000$

ANSWERS FOR CHAPTER 6 REVIEW

1. -9 **2.** -2 **3.** $+6$ **4.** -10 **5.** $+4$ **6.** -3

ANSWERS FOR CHAPTER 7 REVIEW

1. (a) $\dfrac{1 \times 4}{3 \times 4} + \dfrac{1 \times 6}{2 \times 6} + \dfrac{3 \times 3}{4 \times 3} = \dfrac{4}{12} + \dfrac{6}{12} + \dfrac{9}{12}$

$$= \dfrac{9}{12}$$

$$= 1\dfrac{7}{12}$$

(b) $\dfrac{1 \times 6}{5 \times 6} + \dfrac{1 \times 15}{2 \times 15} + \dfrac{2 \times 10}{3 \times 10} = \dfrac{6}{30} + \dfrac{15}{30} + \dfrac{20}{30}$

$$= \dfrac{41}{30}$$

$$= 1\dfrac{11}{30}$$

2. (a) $\dfrac{2 \times 3}{5 \times 3} - \dfrac{1 \times 5}{3 \times 5} = \dfrac{6}{15} - \dfrac{5}{15}$

$$= \dfrac{1}{15}$$

(b) $\dfrac{3 \times 3}{4 \times 3} - \dfrac{2 \times 4}{3 \times 4} = \dfrac{9}{12} - \dfrac{8}{12}$

$$= \dfrac{1}{12}$$

3. (a) $\dfrac{1}{4} \times \dfrac{3}{5} = \dfrac{3}{20}$

(b) $\dfrac{2}{3} \times \dfrac{2}{5} = \dfrac{4}{15}$

4. $\dfrac{1}{8} \times \dfrac{4}{1} = \dfrac{4}{8} = \dfrac{1}{2}$

5. $\dfrac{3}{8} \times \dfrac{7}{1} = \dfrac{21}{8} = 2\dfrac{5}{8}$

6. $\dfrac{4}{21} - \dfrac{2}{15} = \dfrac{4}{7 \times 3} - \dfrac{2}{5 \times 3} = \dfrac{4}{7 \times 3} \times \dfrac{5}{5} - \dfrac{2}{5 \times 3} \times \dfrac{7}{7} = \dfrac{20}{105} - \dfrac{14}{105} =$

$\dfrac{6}{105} = \dfrac{2 \times \cancel{3}}{35 \times \cancel{3}} = \dfrac{2}{35}$

7. $8\dfrac{3}{5} + 9\dfrac{7}{8} = 8 + 9 + \dfrac{3}{5} + \dfrac{7}{8} = 17 + \dfrac{3}{5} \times \dfrac{8}{8} + \dfrac{7}{8} \times \dfrac{5}{5} = 17 + \dfrac{24}{40} + \dfrac{35}{40} =$

$17 + \dfrac{59}{40} = 17 + \dfrac{40}{40} + \dfrac{19}{40} = 17 + 1 + \dfrac{19}{40} = 18\dfrac{19}{40}$

ANSWERS FOR CHAPTER 8 REVIEW

1.
$$
\begin{array}{r}
1.09 \\
\times\ 8.6 \\
\hline
654 \\
8\,72 \\
\hline
9.374
\end{array}
$$

2.
$$
\begin{array}{r}
14.562 \\
\times\ 4.225 \\
\hline
72810 \\
29124 \\
2\,9124 \\
58\,248 \\
\hline
61.524450
\end{array}
$$

3. $8.96 \overline{)3.57} = 896 \overline{)357.00}$
$$
\begin{array}{r}
.39 = 4. \\
\text{x} \\
-268\,8 \\
\hline
88\,20 \\
-80\,64 \\
\hline
\end{array}
$$

4. $1.53 \overline{)4.02} = 153 \overline{)402.00}$
$$
\begin{array}{r}
2.6 = 2.6 \\
\text{xx} \\
-306 \\
\hline
96\,0 \\
-91\,8 \\
\hline
4\,20
\end{array}
$$

5.
$$
\begin{array}{r}
{}^1\quad{}^1\ \\
38.53 \\
42.19 \\
85.12 \\
\hline
165.84
\end{array}
\qquad
\begin{array}{r}
{}^{4\ \ 1}7 \\
16\cancel{5}.\cancel{8}^{1}4 \\
-2\,2.9\ 5 \\
\hline
14\,2.8\ 9
\end{array}
$$

ANSWERS FOR CHAPTER 9 REVIEW

1. $\dfrac{5}{8} = 8 \overline{)5.0^2 0^4 0}$ with quotient $.6\ 2\ 5$

2. $\dfrac{3}{4} = 4 \overline{)3.0^2 0}$ with quotient $.7\ 5$

3. $.6 = \dfrac{.60}{1} = \dfrac{60}{100} = \dfrac{6}{10} = \dfrac{3}{5}$

4. $.45 = \dfrac{.45}{1} = \dfrac{45}{100} = \dfrac{9}{20}$

5. $9\overline{)\,2.0^20^20^20^20^20 \ldots}$ $0.2\,2\,2\,2\,2 \quad \ldots$

6. $0.272727 \ldots = 0.\overline{27} = \dfrac{27}{99} = \dfrac{3}{11}$

ANSWERS FOR CHAPTER 10 REVIEW

1. 12:1

2. 1:17

3. $8 : 5 = 40 : x$
$8x = 200$
$x = 25$

4. $20x = 60$
$x = 3$

5. $300 : \dfrac{25}{4} = 700 : x$

$300x = 25 \times 175$

$300x = 4{,}375$

$x = 14.58 \;\; \text{hours}$

$300\overline{)\,4375} = 3\overline{)\,4^13.^17^25}$ $1\,4.\;5\,8$

See frame 2.

ANSWERS FOR CHAPTER 11 REVIEW

1. $\dfrac{17}{1} \times \dfrac{57}{4} = \dfrac{969}{4}$

$= 242\dfrac{1}{4} \;\text{minutes}$

$= 4 \;\text{hours}, 2 \;\text{minutes}, 15 \;\text{seconds}$

2. $\dfrac{2}{3} \times \dfrac{3}{5} = \dfrac{6}{15} = \dfrac{2}{5}$

3. $\$6.25\overline{)\,\$580} = 625\overline{)\,58000}$

$= 125\overline{)\,11600}$

$= 25\overline{)\,2320}$

$= 5\overline{)\,46^14.^40} \quad 92.8$

$= 92.8 \;\text{ounces}$

ANSWERS FOR CHAPTER 12 REVIEW

1. 40 square feet

2. 25 square feet

3. 8,000 square feet × \$200 = \$1,600,000

4. 900 square feet = 100 square yards × \$20 = \$2,000

5. 80 feet + 300 feet = 380 feet × \$15 = \$5,700

6. 80 feet

7. 30 square inches

ANSWERS FOR CHAPTER 13 REVIEW

1. $\dfrac{220}{7}$ inches = $31\dfrac{3}{7}$ inches

2.   15 feet = πD

$15 \text{ feet} = \dfrac{22}{7} \times D$

$105 \text{ feet} = 22D$

$4.8 \text{ feet} = D$

$$
\begin{array}{r}
4.77 \\
22\overline{)105.00} \\
\text{xx} \\
-88 \\
\hline
17\ 0 \\
-15\ 4 \\
\hline
1\ 60
\end{array}
$$

3. $25 \times \dfrac{22}{7} = \dfrac{550}{7}$ $7\overline{)55^60}\;^{78\frac{4}{7}}$

$= 78\dfrac{4}{7}$ square inches

4. $78\dfrac{4}{7}$ square feet

ANSWERS FOR CHAPTER 14 REVIEW

1. 17%

2. 82%

3. $8 \overline{)5.0^20^40}$ over which $.625$ $= 62.5\%$

4. $\dfrac{5 \text{ inches}}{64 \text{ inches}}$ $64 \overline{)5.000}$ with $.078$ $= 7.8\%$

$$
\begin{array}{r}
x \\
-4\,48 \\
\hline
520 \\
-512 \\
\hline
8
\end{array}
$$

5. 300%

6.

freshmen	100	29.4%
sophomores	90	26.5
juniors	80	23.5
seniors	70	20.6
	340	100.0%

freshmen : $\dfrac{100}{340} = \dfrac{10}{34} = \dfrac{5}{17}$

$$
17 \overline{)5.000} \quad .294 = 29.4\%
$$
$$
\begin{array}{r}
xxx \\
-3\,4 \\
\hline
1\,60 \\
-1\,53 \\
\hline
70
\end{array}
$$

sophomores : $\dfrac{90}{340} = \dfrac{9}{34}$

$$
34 \overline{)9.000} \quad .264 = 26.5\%
$$
$$
\begin{array}{r}
xx \\
-6\,8 \\
\hline
2\,20 \\
-2\,04 \\
\hline
160 \\
-136 \\
\hline
24
\end{array}
$$

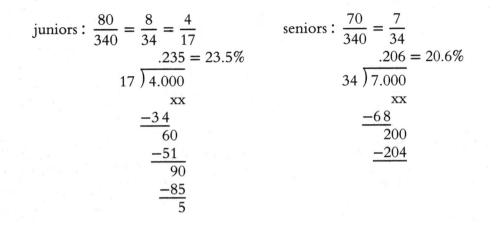

juniors : $\dfrac{80}{340} = \dfrac{8}{34} = \dfrac{4}{17}$

$$\begin{array}{r} .235 = 23.5\% \\ 17 \overline{)4.000} \\ \text{xx} \\ -3\,4 \\ \hline 60 \\ -51 \\ \hline 90 \\ -85 \\ \hline 5 \end{array}$$

seniors : $\dfrac{70}{340} = \dfrac{7}{34}$

$$\begin{array}{r} .206 = 20.6\% \\ 34 \overline{)7.000} \\ \text{xx} \\ -6\,8 \\ \hline 200 \\ -204 \end{array}$$

7. Ten percent of $44.80 is $4.48, and one-half of $4.48 is $2.24. Your tip would be $6.72 ($4.48 + $2.24).

ANSWERS FOR CHAPTER 15 REVIEW

1. (a) $x = 5$

(b) $x = 11$

2. (a) $x = 6\dfrac{2}{3}$

(b) $x = 5$

3. (a) $x = 15$

(b) $x = \dfrac{2}{1.5} = \dfrac{4}{3} = 1\dfrac{1}{3}$

4. (a) $x = -7$

(b) $x = -1$

5. (a) $x = 2$

(b) $\dfrac{3}{4}x = 3$

$x = 4$

(c) $4.2x = -6$
$42x = -60$
$21x = -30$
$7x = -10$
$x = -1\dfrac{3}{7}$

ANSWERS FOR CHAPTER 16 REVIEW

1. 4

2. $5 \times 5 \times 5 = 125$

3. $2 \times 2 \times 2 \times 2 = 16$

4. 9

5. 7

6. $x = \pm 4$

7. $x = \pm\,10$

8. $2x^2 = 18$
$x^2 = 9$
$x = \pm 3$

9. $3x^2 = 48$
$x^2 = 16$
$x = \pm 4$

10. $160 - 81 + 27 = 106$

11. $(4 \times 64) + 24 - 190 = 90$

ANSWERS FOR CHAPTER 17 REVIEW

1. (a) 736,200

(b) 43,000,000

2. (a) nine hundred eighty-seven million, one hundred thousand

(b) one hundred eighty-nine thousand

3. (a) 4,000,000,000,000

(b) 2,000,000,000

4. (a) seven trillion

(b) nine hundred eighty-three billion

5. 8,766,100,000

6. 826,000,000

7. 890,000

8. 45,000,000

9. (a) $8.63 \times 10^8 = 8.\rightarrow 63000000 = 863.000,000$
$\underset{\text{8 places}}{}$

(b) $9.15 \times 10^{-4} = \underset{\text{4 places}}{\underline{0009}}.15 = 0.000915$

10. $8 \times 10^7 \times 9 \times 10^3 \times 4 \times 10^{-12} = 72 \times 10^{10} \times 4 \times 10^{-12} = 288 \times 10^{-2} = 2\underset{\text{2 places}}{\underline{88}}.0 = 2.88$

11. (a) $\dfrac{8.46 \times 10^{15}}{4 \times 10^{12}} = \dfrac{8.46}{4} \times 10^{15-12} = 2.115 \times 10^3 = 2{,}115$

(b) $\dfrac{9.63 \times 10^{-11}}{3 \times 10^{-9}} = \dfrac{9.63}{3} \times 10^{-11-(-9)} = 3.21 \times 10^{-11+9} = 3.21 \times 10^{-2} = 0.0321$ by 3×10^{-9}

ANSWERS FOR CHAPTER 18 REVIEW

1. Let $x =$ Arlene's age

Let $4x - 3 =$ Janice's age

$x + 4x - 3 = 97$

$5x - 3 = 97$

$5x = 100$

$x = 20$

$4x - 3 = 77$

2. Let $x =$ Jason's age now

Let $5x =$ Ron's age now

$5x + 5 = 3(x + 5)$

$5x + 5 = 3x + 15$

$5x = 3x + 10$

$2x = 10$

$x = 5$

$5x = 25$

3.   Let x = first number

Let $x + 1$ = second number

Let $x + 2$ = third number

$$3x = 48$$

$$x = 16$$

$$x + 1 = 17$$

$$x + 2 = 18$$

4.   Let x = first number

Let $3x$ = second number

Let $3x + 2$ = third number

$$7x + 2 = 72$$

$$7x = 70$$

$$x = 10$$

$$3x = 30$$

$$3x + 2 = 32$$

5.   Let x = pounds of peanuts

Let $50 - x$ = pounds of cashews

value of peanuts = $\$2x$

value of cashews = $\$8(50 - x) = 400 - \$8x$

value of the entire mixture = $\$6 \times 50 = \$300$

$$\$2x + \$400 - 8x = \$300$$

$$\$400 - \$6x = \$300$$

$$-\$6x = -\$100$$

$$\$6x = \$100$$

$$x = 16\frac{2}{3} \text{ pounds}$$

$$50 - x = 33\frac{1}{3} \text{ pounds}$$

ANSWERS FOR CHAPTER 19 REVIEW

1. $\$1,000 \times .03 = \$30$

2. $\$2,000 \times 2(.07) = \$2,000 \times .14 = \$280$

3. $A = P(1 + r)^t$
$\quad = \$10,000(1 + .03)^3$
$\quad = \$10,000(1.03)^3$
$\quad = \$10,000 \times 1.092727$
$\quad = \$10,927.27$

$$\begin{array}{r} 1.03 \\ \times\ 1.03 \\ \hline 309 \\ 1030 \\ \hline 10609 \end{array}$$

$$\begin{array}{r} 10609 \\ \times\ 1.03 \\ \hline 31827 \\ 1\ 06090 \\ \hline 1.092727 \end{array}$$

4. $A = P(1 + r)^t$
$\quad = \$4,000(1 + .02)^4$
$\quad = \$4,000(1.02)^4$
$\quad = \$4,000 \times 1.08243216$
$\quad = \$4,329.73$

$$\begin{array}{r} 1.02 \\ \times\ 1.02 \\ \hline 204 \\ 1\ 020 \\ \hline 1.0404 \end{array}$$

$$\begin{array}{r} 1.0404 \\ \times\ 1.0404 \\ \hline 41616 \\ 416160 \\ 1\ 04040 \\ \hline 1.08243216 \end{array}$$

$$\begin{array}{r} 1.08243216 \\ \times\ 4000 \\ \hline 4329.72864 \end{array}$$

5. 10 years

6. 70 years

ANSWERS FOR CHAPTER 20 REVIEW

1. $4.5 \times 500 = 2,250$ miles

2. $6 + 5 = 11$ miles

3. $\dfrac{3,000}{5.5} = \dfrac{6,000}{11} = 545.5$ m.p.h.

4. $\dfrac{80}{1\frac{3}{4}} = \dfrac{80}{1} \div \dfrac{7}{4}$

$\quad\quad = \dfrac{80}{1} \times \dfrac{4}{7}$

$\quad\quad = \dfrac{320}{7}$

$\quad\quad = 45\dfrac{5}{7}$ m.p.h.

5. $\dfrac{460}{115} = 4$ hours

6. $\dfrac{16}{3\frac{1}{2}} = \dfrac{16}{1} \div \dfrac{7}{2}$

$\quad\quad = \dfrac{16}{1} \times \dfrac{2}{7}$

$\quad\quad = \dfrac{32}{7} = 4\dfrac{4}{7}$ hours

ANSWERS FOR CHAPTER 21 REVIEW

1. $\dfrac{340}{969} = 35.1\%$

$$
\begin{array}{r}
.351 \\
969 \overline{)340.000} \\
\text{xx} \\
-290\ 7 \\
\hline
49\ 30 \\
-48\ 45 \\
\hline
850
\end{array}
$$

2. $\dfrac{\$1,500}{\$20,000} = 7.5\%$

3. $\$600 \times .05 = \$30$

4. Let $x =$ original price
$$1.07x = \$428$$
$$x = \frac{\$428}{1.07}$$
$$x = \$400$$

5. $\dfrac{16}{12} = 1\dfrac{4}{12} = 1.33\%$

$$
\begin{array}{r}
2500 \\
\times\ .0133 \\
\hline
7500 \\
7\ 500 \\
25\ 00 \\
\hline
\$33.2500 = \$33.25
\end{array}
$$

6. $\dfrac{\$50}{\$4000} = \dfrac{5}{400}$
$$= \dfrac{1}{80}$$
$$= 1.25\% \text{ monthly}$$

$$
\begin{array}{r}
1.25 \\
\times\ 12 \\
\hline
250 \\
125 \\
\hline
15.00\% \text{ annually}
\end{array}
$$

ANSWERS FOR CHAPTER 22 REVIEW

1. $7,430 + $4,400 = $11,830

2.
$$\begin{array}{r} \$900 \\ \times\ 12 \\ \hline 1800 \\ 900 \\ \hline \$10800 \end{array} \qquad \begin{array}{r} \$10800 \\ \times\ .075 \\ \hline 54000 \\ 756000 \\ \hline \$810000 \end{array}$$

3. Cost = $80 × 100 = $8,000

$$\frac{\$6,000}{\$8,000} = \frac{3}{4} = 75\%$$

4. $59.95
$$\begin{array}{r} \times .6 \\ \hline \$35.970 \end{array} \times 100 = \$3,597$$

5. (a) 150 × $20 = $3,000
$$\begin{array}{r} .97 \\ \times\ \$3000 \\ \hline \$2,910.00 \end{array}$$

(b) 1,500 × $20 = $30,000
$$\begin{array}{r} .93 \\ \times\ \$30000 \\ \hline \$27,900.00 \end{array}$$

6. (a) Let x = annual rate of interest

2%: 20 = x: 360

720% = 20x

72% = 2x

36% = x

(b) Let x = annual rate of interest

3%: 40 = x: 360

1,080% = 40x

108% = 4x

27% = x

7. 1,000 × $40 × .6 × .95 × .98 = $40,000 × .5586 = $22,344

8. Profit = sales − cost = \$1,000,000,000 − \$900,000,000 = \$100,000.000

Profit as a percentage of sales

$$= \frac{\text{profit}}{\text{sales}}$$

$$= \frac{\$100,\cancel{000,000}}{\$1,000,\cancel{000,000}}$$

$$= \frac{1}{10}$$

$$= 10\%$$

Profit as a percentage of investment

$$= \frac{\text{profit}}{\text{investment}}$$

$$= \frac{\$100,000,000}{\$400,000,000}$$

$$= \frac{1}{4}$$

$$= 25\%$$

ANSWERS FOR CHAPTER 23 REVIEW

1. mean = 7, median = 6, mode = 1

2. mean = 8, median = 6.5, mode = 2,7

3. Minimum = 60, Q1= 62, Median = 67.5, Q3 = 71, Maximum 99

4. IQR = Q3 − Q1 = 71 − 62 = 11

5. 1.5 IQR = 16.5. Q1 − 16.5 = 62 − 16.5 = 45.5. No values fall below 45.5, so there are no outliers on the low end. Q3 + 16.5 = 71 + 16.5 = 87.5. Two values, 91 and 99, are greater than 87.5 and can be considered outliers.

6. The candidate's final percentage of the vote would be expected to fall between 46 − 3 = 43% and 46 + 3 = 49%.

WHERE TO FROM HERE?

For most readers, enough is enough! You've relearned most of the math you've forgotten since you got out of school. And as we've said, if you don't use it, you'll lose it. Try to keep using as much as you can of what we've covered here in your daily life, and make wise choices about when and how to use your calculator.

But suppose that you'd like to apply your newly reacquired skills to the study of more advanced math. What I would suggest is that you begin with algebra. We've covered a good bit of algebra in this book, but you can build on that and keep going. Two recommended books are *Quick Algebra Review*, 2nd edition (Wiley) and *Practical Algebra*, 3rd edition (Wiley), both by Peter Selby and Steve Slavin. A more detailed book that does a good job is Douglas Downing's *Algebra the Easy Way* (Barrons).

You might be interested in moving in a somewhat different direction. We've had just a smattering of statistics (in Chapter 23), a field that has a wealth of practical applications. If you're interested in this field, the recommended books would be Donald Koosis's *Statistics*, 4th edition (Wiley), and *Chances Are: The Only Statistics Book You'll Ever Need* (University Press of America). If you'd like to continue studying business math, which we covered in Chapter 22, Business Math, see *Quick Business Math* (Wiley) by Steve Slavin.

Another way you might want to go is to take a formal course in either algebra or statistics. Every college offers these courses, and they are almost always available in adult education programs.

Index

Page numbers in italic refer to problem answers.